"*Nestucca* should be required reading
for anyone involved in oil spill response or recovery.
It reminds us, in intricate detail, of the horrors
and challenges of dealing with a major oil spill....
This type of disaster is still all too possible along
our ecologically fragile coastlines. Time can make us
complacent; thanks to Mr. Webster,
we will not forget."

- Diane Schwickerath
Grays Harbor Audubon Society

"David Webster's *Nestucca* not only lets one experience
the enthralling tales of the dedicated volunteers engaged
in a desperate struggle to save the oil-soaked birds,
but also to listen in on the conversations of
government agencies, each with their own agenda
and personalities, as they try to
work together to prevent disaster."

- Bob Morse, author / birder
A Birder's Guide to Ocean Shores, Washington
Grays Harbor Christmas Bird Count compiler

Nestucca:

an oil spill
turns
creative

Nestucca:
an oil spill turns creative

by David C. Webster
and many other participants

Malamalama Press

Front cover photos:
Dwarfed by rocky Olympic seamonts at low tide, beach walker
provides bright color in foggy scene preserved by Ruth Stennett.

Frightened and helpless, oiled grebe awaits imminent capture
on beach. (Photo by Larry Workman, QIN)

Back cover photo:
On Quinalt Indian Nation beach, Vicki Watkins circles
an oiled murre. (Photo by by Larry Workman)

Cover design by David Webster,
David Starke & Colleen Wasner

ISBN 0-9654023-5-5
Library of Congress Catalog Card No. 99-96154

*10% of net proceeds from the sale of this book
will be donated to non-profit environmental organizations.*

edited by

David Starke
and
Colleen Wasner

Table of Contents

Illustrations

- Map of Washington State Coast, Columbia River to Cape Flattery (Fig. 1, p. 7)
- Map of Northwest Coast, Grays Harbor to end of Vancouver Island (Fig. 2, p. 8)
- Schematic diagram of tug-barge connection (Fig. 3, p. 19)
- Towing vessel *Ocean Service*, plan and elevation views (Fig 4, p. 20)
- Chart of Grays Harbor entrance (Fig. 5, p. 23)
- Drawing of Orville Hook (Fig. 6, p. 24)
- Cape Flattery Geographic Response Plan (Fig. 7, p. 251)
- Clallam Bay to Pillar Point GRP (Fig. 8, p. 252)

* * *

Diane Harvester diary, 56 pages, reproduced in Chapter Five (courtesy Washington State Department of Ecology)

* * *

Black & White Photographs

- Crying Lady Rock
- Bird encased in oil
- Cape Flattery
- State officers inspect beach
- Murres caged at bird center
- Tuber and holder feed murre
- Sudsy scrubbing for oiled murre
- Pacific Rim National Park's Long Beach, Vancouver Island, B.C.
- Dave LeBlanc of Tofino, B.C. (Chapter Nine)

Color Photographs

- Barge *Nestucca*
- *Ocean Service* at Coos Bay, Oregon
- Workers repairing *Ocean Service*
- Cleanup workers on beach
- Helicopter carrying oiled debris
- Clean murres in pen
- Bellingham, Washington pipeline explosion
- Sunset at LaPush

Crying Lady Rock

Weeping mother off this shore,
Stay your tears.

Though greed has raped our heritage,
Inspire us to work together.

In concert may we save, repair, restore
The gifts of nature...

Lest all might weep with you
Forevermore.

- DCW
Summer, 1999

The *Nestucca* Story

Shortly before midnight, December 22, 1988, the tow cable parted between the tug *Ocean Service* and barge *Nestucca*, approaching Grays Harbor on the Washington State coast. As the oil-laden barge drifted toward destruction on the north jetty, the tug skipper quit rigging his Orville Hook recovery device and hurried to put crewmen aboard the barge to pass a line. In doing so, the tug was thrown against the barge's steel hull, ripping a gash through which flowed nearly a quarter million gallons of oil. Despite collision damage to the tug's steering, the leaking barge was recovered and towed offshore. But spilled oil drifted 430 miles (700 km) northwest, fouling beaches of two national parks, tribal lands, and other remote areas. Thousands of marine birds were oiled, and major rescue and shoreline cleanup efforts ensued. This regional event resulted in multi-million dollar legal settlements between Sause Brothers Ocean Towing and various government agencies, creating catalyst funds for new programs dealing with oil spills. International and regional cooperation, including planning, training, equipping and financial recovery regulations for subsequent spills, have grown out of *Nestucca*. It brought the consequences of spilled oil to public awareness in the Pacific Northwest just before the major *Valdez* oil spill in Alaska. The post-*Nestucca* examples of cooperative effort might be replicated anywhere by government, industry, and concerned citizens working together.

Preface

How Did This Happen?

On a January day in 1989 at the Ocean Shores "Dirty Bird Hospital," I spoke with Alice Berkner from the International Bird Rescue & Research Center. Told that the *Nestucca* story might make a book, Alice responded "Go for it!"

Since that occasion, other oil spills and many related developments have lengthened and complicated this story. Even as I write these words, there are remnant loose ends and ongoing events, but....

Why did I start? Perhaps it goes back to the boyhood assignment to clean our chicken house. Necessary but undesirable jobs can bring benefits such as fresh eggs and drumsticks, even fertilizer to help the garden grow. Bad events similarly may bring about good results.

For example, there is a special quality that develops among a group working for a common goal, such as caring for oiled birds. It's also evident in such community projects as soup kitchens or Habitat for Humanity, and I have experienced it in religious settings. Many of us harbor a hunger that responds to need in such a manner that we ourselves are fed or cleansed or housed--as those who've shared such experiences will testify.

As I've moved one step at a time through an unmapped maze, the most important factor has been the cooperation of unnumbered individuals who have responded helpfully. Those interviewed, most with my portable tape recorder, include Captain Raymond Freel of Astoria; Thom Davis of Global Diving & Salvage, Seattle; Chuck Janda of Port Angeles, retired chief ranger of Olympic National Park; Alice Berkner at her home in Vancouver, Washington; Ron Holcomb, Jim Oberlander, Diane Harvester and Dick Logan of Washington State Ecology; Leni Oman and Sara LaBorde of Washington State Fish & Wildlife; Lewey Kittle of the Washington State Health Department, formerly with Ecology; Roland Miller, CleanSound executive; Dr. Larry Galt of the NOAA HazMat staff in Seattle; Denise Daley, Bobby Rose and Vince Cooke of the Makah Tribe at Neah Bay;

Dave LeBlanc of Tofino, B.C.; and Larry Pokeda, officer in charge of Canada Coast Guard's vessel traffic control station at Amphitrite Point, Ucluelet, B.C. One person, Diane Harvester, also contributed the notes from her winter weeks on Olympic National Park beaches.

Several reporters have written portions of this volume: newspaper articles published after the *Nestucca* spill or, in the case of Constance Perenyi of Seattle, in an ornithological journal. Capable writers all, they deserve mention: Stephen Clutter of the *Seattle Times*; Bryn Boerse, Juli Bergstrom and David Wasson of the *Aberdeen Daily World*; Kevin Patterson of the *Peninsula Daily News*, Port Angeles; John Marshall, James Wallace and Lisa Schnellinger of the *Seattle Post-Intelligencer*; Karen Reed of *The Herald* in Everett; John Dodge of *The Olympian*; Jacqueline Storm of Quinault Natural Resources; Theresa Willeford of the weekly *Willapa Harbor Herald* and William R. (Randy) Thomas writing for *Monday Magazine*, Victoria, B.C. The efforts of reporters depend also on the labor of editors such as John Hughes of the *Daily World*, who has been generously helpful.

Holcomb's bound *Summary of Events, News Coverage of the 1988/1989 Grays Harbor Oil Spill*, was my starting point, along with the *Nestucca* Oil Spill On-Scene Coordinator's Report, both from State Ecology. An important contribution came through the Washington State Attorney General, specifically Assistant AG Bill Frymire, who obtained a copy of the 1,200 pages of court depositions by crewmen of the tug and Sause officials. This was matched by the U.S. Coast Guard in providing the report by Lt. Mike L. Emge, investigating officer in the collision between the tug and her barge. With a similar report from Coast Guard Canada and other publications and reports, the remainder of this story has been constructed. There was also the liability trial before Judge James A. Reddin in U.S. District Court, Portland, Oregon.

Numerous other people have contributed, some identified but most not, such as the helpful folk at Thirteenth Coast Guard District Headquarters in Seattle. Others include Larry Workman, Quinault tribal forester; Don Kane, formerly with U.S. Fish & Wildlife; and Ulrich Wilson of the USF&WS coastal refuges office in Port Angeles; Vincent E. Cooke, hazardous-materials coordinator for the Makah Tribal Council, Neah Bay (since in a similar capacity for the Quilleute Tribe); and Carson Boysen of the Northwest Indian Fisheries Commission. Willie Grindstaff of Olympia, retired master cartographer, prepared maps and diagrams from my preliminary drawings and tracings. More could be named, and there was the cooperation of individuals in Port Alberni, Victoria and Vancouver, B.C., and even Toronto, Canada.

Special thanks is due David Starke and Colleen Wasner of Malamalama Press, who initiated contact with me through Black Hills Audubon Society. Their enthusiastic cooperation and experience made possible publication of this manuscript after years of frustration seeking a publisher or financial grant to get this portion of our history available in print.

On a family basis, there is the loving support of my wife, Mary, through years when completing this project seemed as distant as setting foot on the moon. Our two sons, John, opinion page editor for the *Spokesman-Review* in Spokane, and Mark of Gig Harbor both gave providential help to this computer novice. Still another Webster, my brother Dan, of Hanover, Indiana, ornithologist and retired biology professor, provided important technical assistance. Without the help of these and many others, you would not have this book in hand. May you learn as much reading it as I have gained in the research and writing.

- *Dave Webster*

Nestucca

1

Precedents

Nurtured beside the sea, one harbors a lifelong interest in sea life and all things maritime, including its birds. As a youngster, I remember lying transfixed by the marine life in a tidepool trapped in a rocky pocket perhaps 300 feet from our front door in Sitka, Alaska.

There were days we boys spent rowing the family skiff, exploring the nearby--and not-so-near--islands during the 1930's. That was more than 15 years before Rachel Carson's book *Edge of the Sea* made tidepools better known, and 25 years before her classic *Silent Spring*.

We fished and dug clams during those depression days when free food from the sea was an important extension to limited income. Watching a greenling take your baited hook in waving kelp 15 or 20 feet below the boat sticks in one's memory, as does the recollection of using a herring dipnet on a long pole to gather sea urchins from the rocky bottom offshore. I boiled out the urchin shells for sale to tourists from ships that visited Sitka during the summer. Rowing home in the evening from a family berry-picking trip to a channel island, every dip of the oars stirred a pool of light in dark water, and even falling drops were sparkling diamonds of bioluminescence. Who needs fireflies?

And then there was the soap-barrel crowsnest hoisted with the help of my buddy, Bob Yaw, into the upper branches of a tall hemlock. Some 75 feet or more high, we lashed that barrel firmly into place, as a lookout above our treehouse. From it we could see St. Lazaria Island more than 20 miles west beside the open ocean, near Cape Edgecumbe. Later, my two brothers and I rowed out to "Bird Island," so called because of its status as a wildlife refuge. There my finger was nipped by a guillemot when I reached into its nest beneath a boulder. And we quickly learned to wear rainhats walking under wheeling swarms of gulls, murres and puffins.

Memories return with a rush when one goes back years later, as I did with my wife Mary in 1982, only to find that it's never the same. Certainly the beach was not lively with a variety of shellfish, crabs and bullheads--presumably little sculpin--or even an occasional eel hiding in the wet beneath rocks and seaweed. Thanks to many years of untreated chemical discharges from the pulp mill a few miles south, with a nod also to the adjacent small boat harbor, that beach of half a century past had become a wet wasteland, nearly devoid of life.

But positive thanks go also to our parents who instilled love and respect for nature's heritage. One result was that my eldest brother started a career as an ornithologist, and used me as assistant oarsman for the family skiff and as "walk-awayster." Since birds don't depend upon the ability to count for their survival, two people can enter a photography blind and only one depart--the walk-awayster-- and the birds will think no one remains inside. Dan was photographing black oystercatchers (*Haematopus bachmani*, a crow-size black shorebird with a bright red beak) for his master's thesis.

Those childhood experiences created an enduring interest that responded powerfully to the call for volunteers to help at Ocean Shores, caring for thousands of oiled seabirds collected following the *Nestucca* spill at Christmastime, 1988. The thought of beaches blackened with oil was bad enough; helpless murres and scoters fouled as a result of society's dependence on oil provided an even stronger pull.

Other ties to the Washington State coastal area had grown during the interval. I'd worked more than 14 years as a reporter and then city editor for the *Aberdeen World* (now simply *The Daily World*) before leaving to take a state job in Olympia. Covering the news of the Twin Harbors area--Grays Harbor and Willapa Harbor--involved in large measure the two major aspects of the local economy: harvesting the sea and the forests.

Retired from regular jobs, Mary and I could easily drive the 75 miles to Ocean Shores and get involved in the effort to rehabilitate oiled birds for return to the wild. Our part was small in the overall project, but the thousands of volunteers working together made possible the rehabilitation of 30 percent of those bedraggled birds. That was a new high for oil spills, in which 10-20 percent had been the norm.

* * *

Now a few items of the unfortunate history that periodically has blackened portions of Washington's beaches.

Winter along the Pacific coast of North America has earned its reputation as a wet and unpleasant season. Normal weather from October to March is a succession of rainy storm fronts moving through, sometimes with scarcely a break in between. Winds are from the southwest, occasionally northwest, rarely from the north or east. Weather fronts move to the east or southeast, driving across the Gulf of Alaska to slosh over either the Alaskan Panhandle and British

Columbia or further south across Washington and Oregon. Snowstorms, I should mention, are seldom even in winter--more often further north, less so further south along the coast.

It's usually gray day after gray day, with rain accompanied by wind, sometimes fog, and only the occasional sun break. Nobody considers it unusual for the sun to hide for a week or more, and rain is so normal that a day or two without becomes notable. Such dismal weather can bring on cabin fever, but is interrupted by moments of brilliant sunshine that brighten the gray mood with visions of spectacular mountains and ocean.

Rainfall west of the Cascade Mountains varies from 40 inches inland to more than 100 inches along the coast. Mountain areas can get 150-200 inches a year.

Most of us living in the Pacific Northwest learn to cope, enduring rain for the most part by avoiding it, staying indoors at home or at work and going from one to the other using our "four-wheeled motorized umbrellas." Standing in the rain waiting for a bus is for the person without a car, and outdoor work the choice of a different lifestyle.

During the 1930s in Sitka, that rainy town in southeast Alaska 850 miles northwest from Seattle, we walked to school wearing black rubberized cloth raincoats and sou'wester rainhats. That was B.S.: before schoolbuses. But things haven't changed much. Children in Astoria, Aberdeen, Tofino or Ketchikan dress much the same, except today's raincoats may be brightly colored plastic.

For business people, particularly those whose jobs involve water transport, winter is for maintenance and vacations. Doing business in competition with the parade of North Pacific storm fronts can be a losing proposition--but not for everyone.

There are businesses that operate in spite of adverse weather: for instance, Sause Brothers Ocean Towing, of Coos Bay, Oregon. Dale Sause, then 42, president of the company, gave a brief history of the company before a battery of attorneys at his deposition hearing October 3, 1990, in Portland. The legal proceedings covered in chapter eleven were a consequence of the events recounted in this book.

"My grandfather was in the (log) rafting and towing business in Tillamook, Oregon, in the 1930s. The...company...started as a corporation in 1946. The brothers were...Curt and Henry, partners in Henry Sause & Sons Rafting & Towing. In the late '40s they were towing logs and developing a business along the West Coast primarily in forest products. They developed the self-dumping log barge system. They modernized the towing and delivery of logs out of the Tillamook

Burn area to Grays Harbor (and) Puget Sound.

"And that...developed...into the coastwise towing business in the early '50s. At that point...there were three brothers involved,...Henry Sause Jr., Curt Sause and Paul Sause. In the '50s, our primary business was the delivery of packaged lumber into Southern California. In the '60s we began to diversify, and our business expanded into...Hawaii and Alaska. In the '70s we continued...into the petroleum and chemical business. At the end of the '70s we developed a shipyard and repair facility. Now our business is basically comprised of our ocean towing division, our stevedoring division which takes care of...loading and discharging...vessels, and our shipyard division."

Question by Michael Haglund, attorney representing the State of Washington: "You mentioned a self-dumping barge. Was that an invention of Sause Brothers during the '40s?"

A: "That is correct, we developed that system. And then it has been copied by the Canadians and several other companies."

Q: "This was a barge that could automatically dump logs or loads of logs that were aboard the barge during...transit?"

A: "Yes. ...And (when) we purchased the *Ocean Service*...we added the stern roller and (retractable) tow pin arrangement, ...something our company developed in the late '70s."

Q: "How many...are employed by Sause Brothers Ocean Towing and its affiliates?"

A: "Approximately 450."

Q: "Now you mentioned that you diversified.... Are those divisions separately incorporated...?"

A: "Yes, they are separate corporations.... (There is) Crescent City Marine Ways & Dry Dock Company, headquartered in Portland, which operates stevedoring activities in Los Angeles, San Francisco, Coos Bay, Longview and Seattle.

"Southern Oregon Marine,...headquartered in Portland, has ...one shipyard in Coos Bay, Oregon. We specialize in tug and barge repair and...building petroleum barges. We have built five barges since 1980.

"We (also) have a...company, Willamette Leasing, and a company that's based in Honolulu,....Sause Brothers, Incorporated."

Q: "And what is its business?"

A: "Inter-island towing and towage in the South Pacific."

Q: "How many tugboats...and barges...does Sause Brothers own?"

A: "Thirty tugboats...and approximately thirty...barges."

Mr. Sause estimated, in the course of his questioning by attorneys preparing for the United States District Court trial, that Sause handles about 80 percent of the winter towing business along the Pacific Coast.

"Every October 1...we start wintertime procedures,....go through extensive inspections. We review all (towing) wires, review maintenance, winches, brakes, fairleads, all the rollers. We review, on the barges, all the shackles, the flounder plates, pigtails, surge gear, bridle legs, towing padeyes. We go through and winterize all equipment, antifreeze, those types of things. We make sure that the barge lashings, (their) hatches, and all hatches on the tugs, everything is ready for wintertime."

In regard to personnel for winter ocean towing, and specifically for the trip to Grays Harbor, Sause explained that the company had a prime skipper and chief engineer operating a prime boat, the tug *Ocean Service* under the command of Charles E. May III (whose father also had been a tug captain) with Jack Wilson as chief engineer. At that time, December of 1988, he said the company had six prime boats and 12 prime skippers, all of whom Sause himself had designated for their skill, experience and dependability.

But, as we all learn eventually, nobody's perfect; and events can conspire against even the most skilled to threaten disaster. Those same skills sometimes can prevent the worst from happening, as was the case with the *Ocean Service* and its barge, the *Nestucca*.

Not that this was the first such oil spill on our coast. The earliest major spill I've been able to identify was in 1964. A 200-foot fuel barge towed by the Seattle tug *Neptune*, carrying 56,000 barrels (2,352,000 gallons) of gasoline, diesel and stove oil, went adrift March 11, 1964, off Grays Harbor (see figure 1, page 7).

Since I was involved as city editor of the *Aberdeen World*, let's flesh out the bare bones of the story as it unfolded.

The newsroom of the *World* at that time was on the second floor of a brick building in downtown Aberdeen, on Grays Harbor about halfway along the Washington coast between the Olympic Peninsula and the Columbia River. About 30 by 40 feet, the newsroom had one corner partitioned off with a window wall as the "wire room" for a battery of clattering teletype machines that delivered news from elsewhere in the larger world.

Four desks were positioned in line by large windows overlooking the street. Closest to the backshop composing room door and farthest from the entrance at the top of the stairs leading from the street was the desk of the managing editor, Edwin Van Syckle. Next were a pair of desks occupied by Ade Frederickson, news editor, and Irv Seath, wire editor. Then, turned 90 degrees so as to face across the newsroom, was the desk where I worked, handling local news copy generated by the three reporters at desks across the room, plus a pair of field reporters, one on Willapa Harbor and the other at Montesano to cover Eastern Grays Harbor County. I had worked on the city desk at the time for some five years.

One reporter desk was staffed by Barbara Elliott, city hall and schools specialist. Next to her worked Wendell Keene, the Hoquiam and north beaches reporter. A third was for the general assignment reporter, Carl Enis, and a fourth for the sports editor, Robbie Peltola.

Van Syckle--and perhaps one or more of the others--has since died, but on that day in 1964 all were on the job, a normal day in the life of a typical small newspaper. But no day in newswork is ever routine. That's one attraction of the work.

Actually it was March 12, 1964, because the accident happened during the night, when a telephone call advised that a fuel barge had drifted ashore between Pacific Beach and Moclips, on the ocean beach some 15 crowfly miles west and almost an equal distance north. Phone calls to Coast Guard Group Grays Harbor at Westport and to others who could provide information helped us develop a story for that day's front page. It was not the top story, however. The conviction of Jimmy Hoffa on conspiracy charges rated that spot. Wendell Keene, with help from others of us on the staff, put together the following account (Endnote 1-1):

"PACIFIC BEACH--Bound for Oregon and California, a 200-foot barge carrying 56,000 barrels of gasoline and diesel oil in its tanks broke adrift yesterday, snapped a salvage hawser last night and drifted onto a sandy shoal between Pacific Beach and Moclips.

"Pounding surf, 8 to 10 feet high and breaking, prevented the Coast Guard from getting lines aboard the barge at high tide today, and salvage attempts had to be called off to await calmer conditions.

"The danger is that the barge might break up and loose her cargo, creating not only the possibility of fire but of extensive damage to clams and other marine life in the area.

BURIED IN SURF

"Surf was breaking over the barge this morning as she rested on the sandbar about 300 to 400 yards offshore at high tide. She had grounded at about a 45-degree angle, according to CWO J.S. Breschini, Group Grays Harbor Coast Guard commander, but later swung broadside to the beach. Occasionally the bow seemed to lift slightly, he said.

"There was a noticeable odor of gasoline, presumably from the tank vents, but no indication of leakage, according to Breschini.

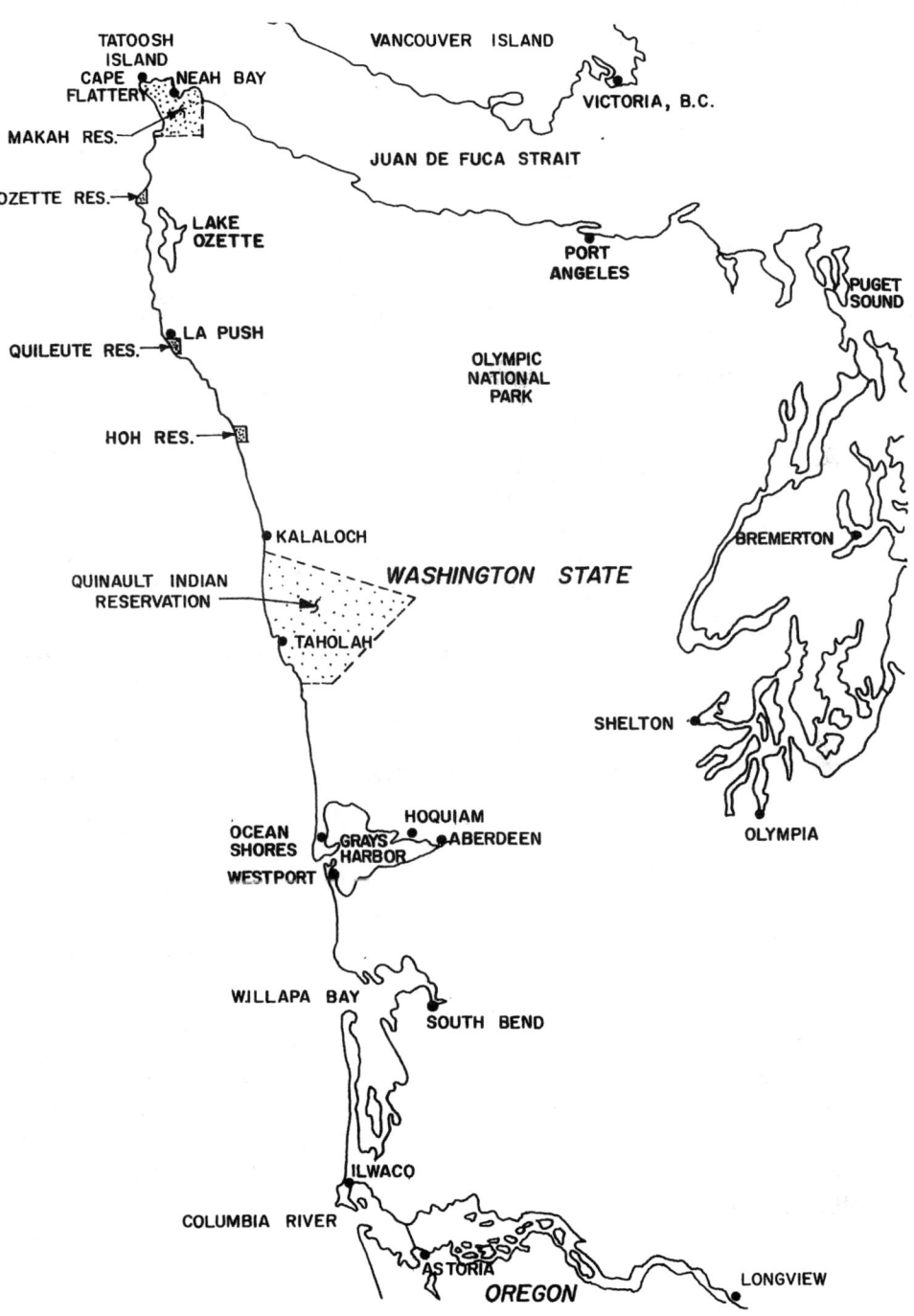

TATOOSH
ISLAND
CAPE ● ●NEAH BAY
FLATTERY

MAKAH RES.

VANCOUVER ISLAND

VICTORIA, B.C.

JUAN DE FUCA STRAIT

OZETTE RES.→

LAKE
OZETTE

PORT
ANGELES

PUGET
SOUND

LA PUSH

QUILEUTE RES.→

OLYMPIC
NATIONAL
PARK

HOH RES.→

● KALALOCH

WASHINGTON STATE

BREMERTON

QUINAULT INDIAN
RESERVATION →

●TAHOLAH

SHELTON

HOQUIAM
OCEAN   ●GRAYS   ●ABERDEEN
SHORES   HARBOR
WESTPORT

OLYMPIA

WILLAPA BAY

SOUTH BEND

ILWACO

COLUMBIA RIVER

ASTORIA
OREGON

LONGVIEW

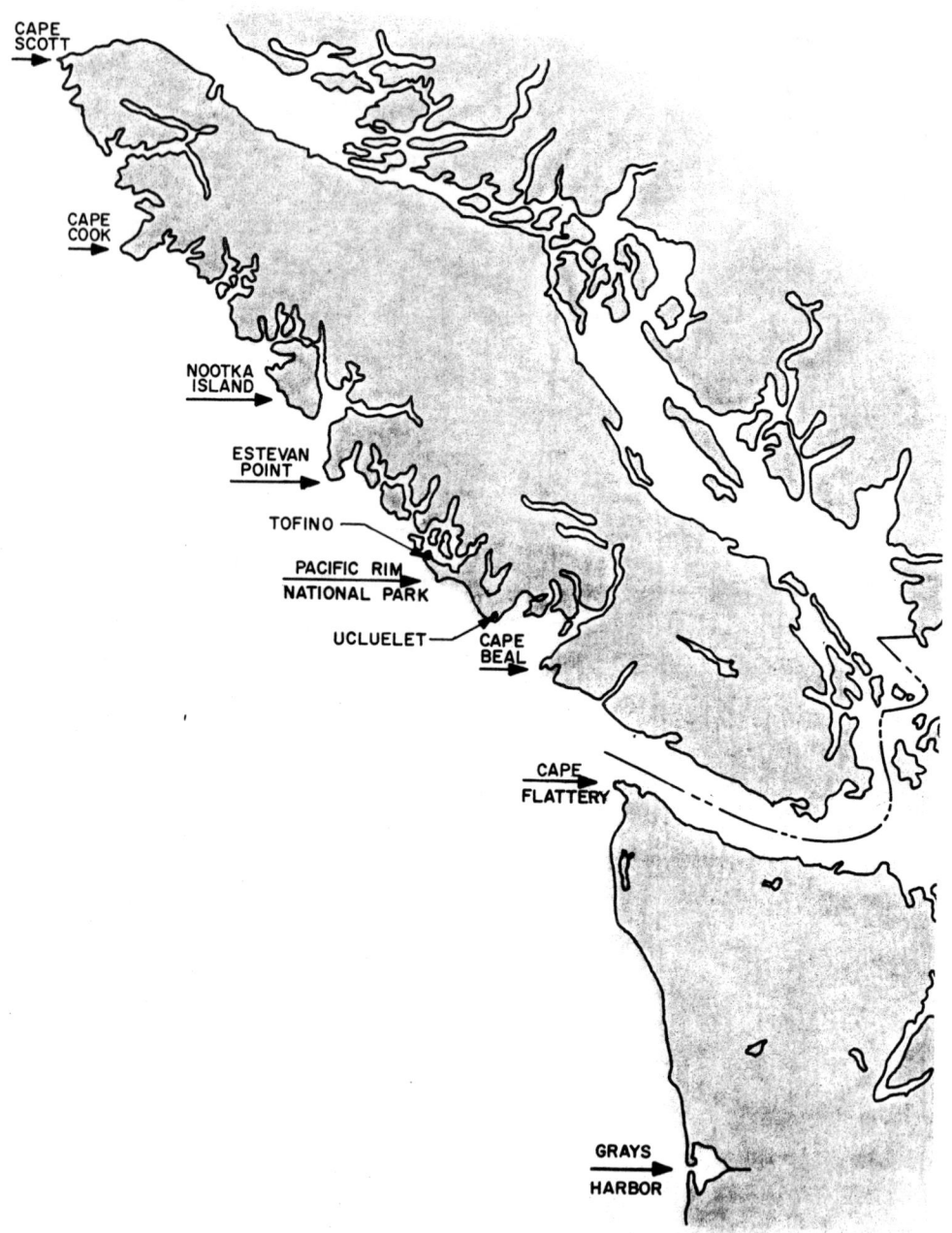

"The $600,000 barge is owned by United Transportation Co. of San Francisco and was loaded at Anacortes and Ferndale refineries for delivery at Coos Bay, Ore., Crescent City and Eureka, Calif.

SNAPS TOWLINE

"Under tow of the Seattle tug *Neptune*, the barge snapped its hawser (sic., see later report from the Coast Guard hearing) in the storm Tuesday night. Located by a Coast Guard plane yesterday, it was taken in tow by the USCG Cutter *Modoc* from Astoria. But the line parted again and the barge drifted into the surf.

"In addition to the cutter, the Coast Guard had a 36-footer (motor rescue boat) from Westport standing offshore this morning, a helicopter from Port Angeles to assist in transferring lines, and beach crews from the Pt. Grenville Loran station and Westport.

"When the salvage attempt was prevented by the rough surf conditions, the Coast Guard recalled its units. Two cranes were reported to be on the beach early this afternoon, and it was possible that attempts to get lines aboard from shore might be tried at low tide."

The story continued a few more paragraphs, quoting J.E. Lasater, assistant state fisheries director at Olympia, expressing concern over the possibility of extensive damage to fish and shellfish in case the barge's cargo spilled.

Keene went to the scene to keep close track of events, and we also sent Bill Jones, Aberdeen photographer, on special assignment. He brought back good photographs of the barge in the surf, which were on the front page the next day. "Turks threaten Cyprus intervention" was the banner headline that afternoon; but the second story, at the top left of page one, was Wendell's report.

"MOCLIPS--North Beach clams still appeared to be safe as salvage preparations went forward this afternoon to pull a 260-foot barge loaded with fuel from (its) parallel perch along the Moclips beach.

"State Pollution Control Commission and Fisheries Department personnel were at the scene, anxiously hopeful the barge's 2 million gallons of gasoline and oil would not be released to kill coastal marine life, especially clams.

"Though a petroleum odor is detected by those arriving on the beach, Al Daugherty, Aberdeen enforcement officer for the Fisheries Department, said the big barge doesn't appear to be leaking.

"The Puget Sound Tug & Barge Co. tug *Sea Witch* was at the Port Dock (between Aberdeen and Hoquiam in Grays Harbor) at noon today to take on 18,000 feet of heavy line. Salvage Master George Mitchell of United Transportation Co., San Francisco, owner of the barge, said the tug would sail at 4 o'clock this afternoon for Moclips.

WORK THROUGH NIGHT

"'We'll work to pull the barge off the sand just as soon as we get there,' Mitchell said. 'We'll work through the night.' He said at least one other tug would help.

"Salvage efforts this morning began by shuttling of three men to the barge with a chartered helicopter operated by a former Hoquiamite, Budd Darling of Chehalis. Their inspection was brief but they reported only slight damage to the pumphouse at the stern and no evidence of holes or cracks."

The story continued, reviewing information from the previous day, and adding that a minor spill from the barge had killed crabs, surf perch and razor clams in the immediate area.

It may have been that day we received a telephone call from a person at the scene who was not involved in the salvage effort. On condition of anonymity, this person said that the barge was indeed leaking large quantities of fuel.

That day's banner headline concerned the conviction of Jack Ruby in Texas for killing Lee Harvey Oswald, the accused assassin of President John F. Kennedy.

Monday evening, March 16, the tug was reported still pulling in a continuing effort to get the grounded barge refloated. The next day's paper finally gave the situation a banner head: "NORTH BEACH SHELLFISH WIPED OUT" and a deck head below, "Clam oil-kill ends digging; barge freed."

"Total kill of all beach and inshore marine life for more than 10 miles along the North Beach was revealed this afternoon by State Fisheries Department experts.

"Ruination of one of Washington State's finest razor clam digging areas followed the salvage early today of a stranded fuel barge from the beach near Moclips with what remained of its cargo of more than 1 million gallons of gasoline and diesel and stove oil.

"'Everything is dying,' said J.E. Lasater, deputy director of fisheries, who termed the situation a major marine disaster. Lasater said the kill included sea birds, fish ducks, tube worms and moon snails."

Extensive damage on beaches of the Quinault Indian Reservation further north also was reported.

The offending fuel barge was pulled from the beach at 1:16 a.m. by a 9-inch hawser from the tug *Sea Witch* standing offshore.

"The *World* learned," continued the story, "that a total of 500,000 gallons of diesel and stove oil was pumped from the stranded barge between 2 p.m. and 11 p.m. yesterday, and that 750,000 gallons, about half of the gasoline aboard (actually 53 percent), had leaked out while the barge was on the beach.

"Earl Coe (Washington Secretary of State), as chairman of the State Pollution Control Commission, said this morning he gave permission yesterday to 'pump out water and oil from the barge bilges' to lighten it enough so it could be pulled free.

"'It was a method of saving the ship as against having it entirely destroyed and all the oil dumped on the beach,' he said."

Mr. Coe died only a month or so later.

Charles Roe, retired from state service as an assistant attorney general, confirmed the accuracy of accounts published by *The Daily World*. Roe said he participated in a three-way phone conversation with Coe and Stanley Knox of the Pollution Control staff. Coe was in Portland, Ore., and Knox on the scene at Moclips, while Roe was in Olympia. He explained that Coe did not know the barge lacked bilges, and nothing separated the oil from the ocean water except a single layer of steel plate. Mr. Coe approved pumping out the fuel "by not turning it down" when it was proposed, and took personal responsibility for the decision when objections were raised later on, said Roe.

It is still true that most fuel barges and tankers in the world are of single hull construction.

The commercial clam digging season was closed immediately, as was sports digging north of the Copalis River, and March 20 fisheries agent Daugherty reported dead clams the full length of the beach south of the spill, along with dead or dying birds that included grebes, scoters and sea pigeons (pigeon guillemots). He could see an oil slick offshore. A North Beach businessman, William Anable, said however that he had found no dead clams on the beach except those that had washed ashore.

At the ensuing Coast Guard hearing it developed that the tug *Neptune*'s 1,400-foot towing cable had run off the winch drum. Capt. Fritzjof Berge testified that he had 1,200 feet of cable out and 200 feet remaining on the drum when he turned the watch over to the mate at midnight. An hour later he was awakened by a surge as the tug lost her tow. No explanation was reported as to why the cable ran off the drum.

In another follow-up story, George Starlund, state director of fisheries, said an estimated 32,000 pounds of razor clams had been killed (1-2). But the clams were not wiped out, and eventually recovered.

State Attorney General John J. O'Connell later that year agreed to an $8,000 payment by United Transportation Company as an out-of-court settlement for damages caused by the spill (1-3).

Another oil spill along the Washington coast, noted only briefly at the time and mostly forgotten since, was the wreck of the unmanned military troop transport *Gen. M.C. Meigs* in January of 1972 (1-4). The *General Meigs* had been moored with the U.S. Maritime Service reserve fleet in Budd Inlet near Olympia at the south end of Puget Sound. When the Olympia reserve fleet was closed out, the *Meigs* was determined to be worth saving rather than being dismantled for scrap. So she was being towed under Navy contract from Olympia to intended mooring with the USMS reserve fleet in San Francisco Bay.

Contending with gale-force winds and 30-foot seas, a tug operated by Murphy Pacific Marine Salvage Co. reported that its towing cable snapped about 2 a.m. January 9, 1972, some nine and a half miles off the Olympic Peninsula. The *Meigs* drifted ashore at the southwest corner of the Makah Indian Reservation, about seven and a half road and hiking miles south of Neah Bay, between Portage Head and Shi Shi Beach. That's about 8 miles southeast from Cape Flattery as the seagull might fly.

After grounding on a rock ledge less than 200 yards out and parallel to the shoreline, the *Meigs* broke apart. Her forward two thirds listed to port, with the bow heading southeast, while the aft section swung more to the north. Eventually the fantail broke away and was swept closer to the beach among the seastack rocks. A large seastack separated the two main sections of the hulk in aerial photos taken by the *Seattle Times* and also photos five years later when the stern section was underwater except for the after mast (1-5).

Commissioned during the 1940s and reactivated during the Korean war, the *Meigs* was decommissioned in 1958 and had been in the reserve fleet since. Navy records indicated the 622-foot, 12,000-ton ship carried 116,690 gallons of fuel oil in 19 tanks ranging from 8,000 to 22,000 gallons capacity. Whether these tanks all were full is doubtful, and in fact the second-day story in the *Seattle Times* said the vessel carried 2,200 barrels of oil, or 92,400 gallons. The latter figure may have been the quantity actually measured in her tanks during storage, as that was routine practice, according to Archie Townsend of Olympia, a former reserve fleet worker.

The bunker fuel (Navy special) burned by such steam-powered ships will flow through a pump only when heated, and at the temperature of seawater is but semiliquid. Even so, a major single oil spill occurred the first two days after the *Meigs* went ashore, and oil continued to appear in the area for five years. Researchers found the heavy oil affected only the vicinity of the rocky cove protected by the wreck and seastacks (op.cit. 1-5).

The largest marine spill in Washington State history--thus far-- was December 21, 1985, when the tanker *Arco Anchorage*, 120,000 deadweight tons, 883 feet long by 138 feet wide, made a bad turn into Port Angeles harbor on the south shore of Juan de Fuca Strait. She cut too close to Ediz Hook, then went too far south and into water too shallow for her 51-foot draft. A rock punctured the inch-and-a- half-thick steel on the bottom of her number five center and portside tanks, releasing 231,000 gallons of Alaskan crude oil. It was comparable in quantity with the later *Nestucca* spill of 227,304 (plus or minus 4,200) gallons (1-6).

Lewey Kittle, former on-scene coordinator for the Washington State Department of Ecology, added some explanatory information. The accident occurred, he recalled, because the pilot put on too much speed so that the deep-laden tanker ran onto the shallower bottom. Then he reversed engines and turned, trying to back the ship into deeper water. The maneuver shoved the vessel over a large buried rock, ripping a gash from near the stern forward through the bottom of tanks from which the crude oil flowed.

Port Angeles harbor was transformed into a pool of oil, and the tide carried some of it eastward 14 miles to Dungeness Spit and its National Wildlife Refuge. The final toll was about 4,000 birds and 11,000 pounds of clams killed. Atlantic Richfield (ARCO) spent $13 million to clean the spill and settle damage claims, winning praise for its fast action and generally good response.

Port Angeles Fire Chief Larry Glenn, recalling the spill, said, "The first thing we noticed was all of a sudden it became very silent." Glenn was in a fire department utility boat as it approached the scene that evening. The boat's wake vanished as it moved into the oil slick, and the water was flat as a floor, Glenn said. Any doubts about the magnitude of the spill were erased the next morning, according to a recollection story five years after (1-7). "It was total disbelief," Glenn said. "It was going bloop, bloop on the shore. That's when I realized what a horrible catastrophe it was."

"ARCO decided that protecting Dungeness Spit was the top priority, and concentrated efforts there," said the AP story. "The cost was to let the oil have its way on Ediz Hook, which juts out and around the Port Angeles harbor. But Dungeness Spit remained generally clear of oil, and scientists agree the environmental legacy of the spill would have been much different if the entire spit were (to have been) coated.

"Bob Levine, ARCO's spill coordinator, now calls protecting (Dungeness) spit the best decision he ever made. On Ediz Hook, the oil hit the beach so quickly that little could be done to prevent damage.... It took until April (more than three months) to clean the beach. "June Siva, ARCO's manager of environmental sciences, said the company conducted tests of sediment samples on the hook for three years. In the first two years, she said, 'hot spots' of oil could be found, but marine life in the inner tidal areas was recovering. 'We saw juvenile clams coming back...' she said, and '...now clams are most likely at pre-spill levels'."

There have been other spills over the years. One non-spill is worthy of mention, especially in contrast to the *Arco Anchorage*, because the two events occurred in such proximity with differing results. The Japanese tanker *Matsukaze* was inbound to Puget Sound in 1988, but according to Kittle of DOE someone laying out the course to Tacoma forgot about the Olympic Peninsula. The plot was established as a line southeasterly from the point where the *Matsukaze* turned into Juan de Fuca Strait. On automatic pilot, she ran onto the beach at her full 14-knot cruising speed some 17 miles west of Port Angeles.

As the bow of the tanker plowed to a stop only 50 feet or so short of the high tide line, a man who had parked his motorhome on the beach heard the prolonged "cru-u-u-unch" and looked out to see the ship. He switched on his CB radio, intending to report the grounding to the Coast Guard, but found he was in a radio "dead spot" between headlands, and could raise only someone in a nearby home. The resident didn't believe his report of a ship on the beach, so he had to find a payphone to notify the Coast Guard. The ship's grounding was so violent that a Volkswagon-size boulder from the beach was shoved through the steel hull to the inside of the tanker, according to Kittle, and yet not a drop of oil was spilled. As you may have guessed, the reason was that the tanker had a double hull. Even that big boulder was not enough to rupture the oil tanks inside.

The difference between the *Arco Anchorage* and *Matsukaze* events, as Eric Nalder observed in a *Seattle Times* series on oil tankers, was $13 million for cleanup of the Port Angeles spill, plus another half million dollars to repair the ship. For the *Matsukaze*, repairs cost $213,000 (1-8).

There now are finally requirements under the federal oil pollution act of 1990 that tankers have double hulls, although it delays the date for total compliance until the year 2015, a variable deadline depending upon the age and size of a vessel. A tanker with a single hull may not operate in U.S. waters after January 1, 2010, except for unloading at a licensed deepwater port or offloading in a lightering zone more than 60 miles offshore (1-9). The growing costs of oil spills has begun to make double hulls look better and better, however, and may bring earlier compliance.

One other spill needs mention, because it involved an oil barge on Puget Sound operating under a lack of proper regulation. In January, 1988, almost a year before *Nestucca*, the *MCN-5*, leased by Olympic Tug & Barge of Seattle, overturned and sank in Guemes Channel only a few hundred yards from the Washington State ferry dock west of Anacortes. It released 67,368 gallons of bunker fuel after sinking before it was carefully raised with the remainder of its 414,708-gallon cargo. Most of the heavy oil that escaped the overloaded barge was swept away by tidal currents, but beaches in the vicinity were affected.

The barge formerly had been licensed for ocean hauling, but was reinspected by the Coast Guard and relicensed for inside waters carrying oil in two end tanks previously reserved for buoyancy. As a result the barge was unstable when deeply loaded, but not in violation of non-existent regulations for protected waters. Substantial financial assessments were lodged against the responsible parties by state and federal authorities (1-10).

Now let's move on to our primary case.

2

Misfortune and Heroics at Sea

The weather at Ferndale, near Bellingham on northern Puget Sound, was typical for winter, gray beneath a cloudy sky, when the Sause Brothers tug *Ocean Service* left the BP (British Petroleum) refinery towing the loaded barge *Nestucca* at 1335 hours (1:35 p.m.) December 21, 1988 (2-1).

The *Ocean Service* is a twin screw, diesel-powered vessel of 3,800 horsepower, 121.1 feet long, 193 gross tons. Her tow, the tank barge *Nestucca*, 5,339 net tons, is 301.8 feet long, 90 feet wide and 21.4 feet deep. The *Nestucca* carried in her single-skin steel hull 70,000 barrels (2,940,000 gallons) of bunker-C fuel oil destined for Aberdeen and Portland. The barge's draft forward was about 17 feet 4 inches, and 18 feet aft, according to the Coast Guard report.

Tug and barge are connected by 2,600 feet of wire rope (cable) 2 1/4 inches in diameter, which in normal towing conditions is--except for about 200 feet retained on the tug's winch drum--stretched in a hanging arc in the water between the two vessels. That towing length is shortened to about 1,500 feet or less in narrow or restricted waters. The towing cable connects to a chain pigtail of 45 feet, another 45 feet of surge chain and twin 60-foot chain bridles attached to the forward corners of the barge (see figure 3, page 19).

There had been a little drizzle the day before, but this was between weather fronts, and Puget Sound is partially protected from the storms that march in succession across the North Pacific during winter, thanks to the mountains on Vancouver Island and on the Olympic Peninsula. Another front was moving in, however, and after the *Ocean Service* traveled westward out Juan de Fuca Strait, Captain Charles E. May III was forced to reduce speed as they slogged south along the Washington coast after making the left turn around Cape Flattery and Tatoosh Island (see figure 1, page 7).

Captain May said in his later deposition to the attorneys (2-2), "The weather had picked up and it was kind of nasty, and then it was laying down, the weather come down in good shape. We were down on turns for awhile on that voyage, I believe, and, like I say, the weather kept getting better, the wind quit blowing and at the tail end of the voyage we were up to full turns."

One should explain a couple of terms here. By "laying down" the skipper refers in particular to the rolling ocean swells and any overriding smaller waves or wind chop. And by "turns" he is referring to the engine revolutions powering the tug's twin propellers.

Had the entire voyage been at or near the possible maximum speed of about eight knots--nine miles per hour--the *Ocean Service* would have approached the Grays Harbor entrance sometime in the afternoon. But as it worked out, the approach was near midnight Dec. 22, which was also close to flood (high) tide, the necessary condition for crossing a bar with a loaded tow.

The Grays Harbor bar is among the most notorious along the Pacific Coast for mariners, perhaps second only to the bar at the entrance to the Columbia River for the manner in which the seas can rise up--literally--to endanger any vessel. Sneaker waves, as they are known, can be as high as 40 feet--a full 12 meters. Once, during the 1960s, the Coast Guard motor lifeboat *Invincible* went out after a commercial crab boat that had been swamped off Westport, and the *Invincible* herself was rolled and immobilized overnight before she could be towed back to port. Mariners have a healthy respect for bar conditions there. During January of 1993 a successor *Invincible* also rolled off Westport, according to a news report.

In addition to Captain May, who had been with Sause Brothers Ocean Towing Company of Coos Bay, Oregon, since 1969, crewmen of the *Ocean Service* that trip were Gary Thomas Rickey, mate; Jack A. Wilson, engineer; Curtis Lee Bartley, deckhand and assistant engineer (a nephew of the skipper), and Cecil Johnson, cook. The crew of a tug usually is on duty for about a month, then is replaced by a second Sause crew. Junior crewmembers may serve longer.

Son of a towboat skipper, May graduated from high school in his hometown of Crescent City, California, and spent three years aboard commercial fishing vessels working coastal waters. He spent three winter months in the woods as a choker setter on a logging crew, then decided "there might be bigger, better things," so he signed on with Sause Brothers in May of 1969. In the ensuing 20 years he worked every towboat job, moving up to master in a couple of years, and has been to the South Pacific, Alaska and the Gulf of Mexico. He had skippered the *Ocean Service* since its acquisition by Sause in 1984 (see figure 4, page 20, and color photos 2 and 3).

Jack Wilson, chief engineer, age 42, had joined the U.S. Navy after high school, served six years and then worked for another towboat firm for five years before signing with Sause Brothers Ocean Towing in 1978. He was licensed as an engineer in 1983, and had a year aboard the *Ocean Service* before December of 1988.

Gary Rickey, the mate, age 32 at the time, had worked on commercial fishing boats off Nova Scotia as a youth. After two years

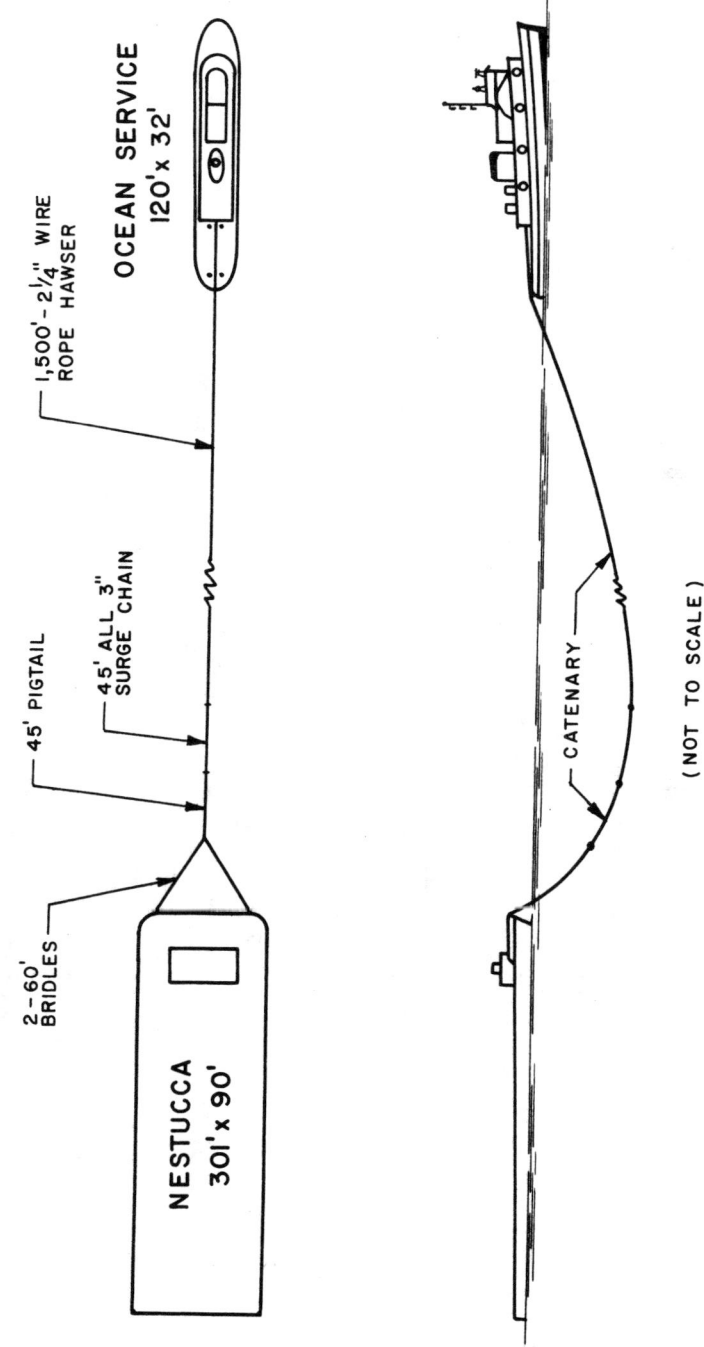

OCEAN SERVICE
120' x 32'

1,500' - 2¼" WIRE
ROPE HAWSER

45' ALL 3"
SURGE CHAIN

45' PIGTAIL

2 - 60'
BRIDLES

NESTUCCA
301' x 90'

CATENARY

(NOT TO SCALE)

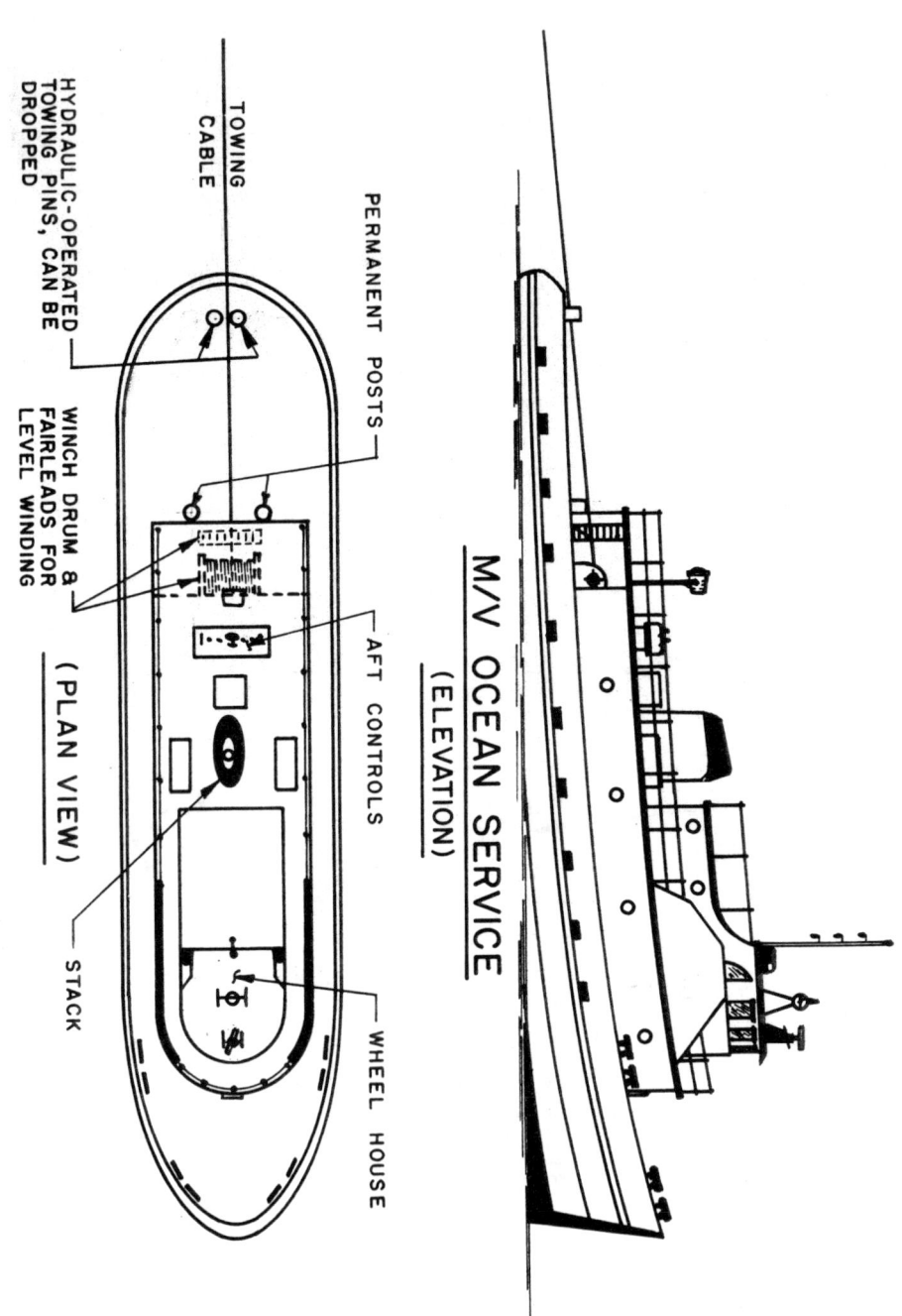

HYDRAULIC-OPERATED TOWING PINS, CAN BE DROPPED

TOWING CABLE

PERMANENT POSTS

WINCH DRUM & FAIRLEADS FOR LEVEL WINDING

AFT CONTROLS

(PLAN VIEW)

STACK

WHEEL HOUSE

M/V OCEAN SERVICE

(ELEVATION)

of college in Massachusetts, he married and moved to Oregon to take up fishing. He left it for two years working towboats in Alaska, then started with Sause in November of 1988, less than two months before this incident.

Bartley had attended welding school and worked as a welder and ranch hand in California, then six years as a logger before taking the job with Sause in November. This was only his second trip as deckhand and engineroom assistant.

The cook on a tug usually is not involved in affairs outside the galley, but Cecil Johnson was pressed into helping rig the Orville hook when Rickey and Bartley were aboard the barge.

Captain May recounted later for the U.S. Coast Guard and also at his deposition that he normally turned in toward Grays Harbor from about the 60-fathom line, which he had followed from Cape Flattery, at a point slightly south of the Point Chehalis Range. This line on the marine chart (see figure 5, page 23) is slightly north of west (253 degrees magnetic) from the Group Grays Harbor Coast Guard station at Westport and parallel with the south jetty. It almost lines up with Buoy 3.

Had he been coming from the south, Captain May would have made the turn at buoy GH, the sea buoy, about 3 miles further south. But from the north he turned at buoy 3, saving about six miles of travel, and this later caused some criticism in a *Seattle Post-Intelligencer* story (2-3). Reporter James Wallace quoted a Grays Harbor pilot as saying, "You don't go that way, not in that kind of weather, especially when you are towing an oil barge. You go down to the sea buoy."

But Lt. Mike Emge, the Coast Guard investigating officer, discounted criticism from Grays Harbor pilots, suggesting that since they would like to have one of their own guiding tugs in and out, any second guessing "should be taken with a grain of salt," according to the *P-I* story.

Harbor pilots are required on foreign-registered commercial vessels in state waters, and on oil tankers over 50,000 tons, but are not required for tugs hauling barges. Capt. Raymond Freel of the tug *Janet R* told Lieutenant Emge he had noticed little or no difference between the two routes earlier.

This *P-I* story was complimentary, suggesting that the skipper and crewmen of the *Ocean Service* deserved praise for their heroics in preventing total loss of the *Nestucca* and eruption of its entire cargo of nearly 3 million gallons of fuel oil compared with the actual spill of less than 1/4 million gallons. That was bad enough, of course, nearly equivalent to the 1985 *Arco Anchorage* spill at Port Angeles.

But we're getting ahead of events.

Captain May was asked at his deposition hearing to recall the details as he remembered them from the approach to Grays Harbor that night.

He described first how he makes a turn in small pieces, so as to nudge the towed barge a little at a time toward the new direction, such as around the seaward buoy where he set his course toward Westport. As to weather and sea conditions, Captain May responded:

"I think, you know, like an eight-foot swell, six to eight-foot swell, fairly stretched out swell, not a sharp swell. No wind, the wind had died off."

Q: "When you say a stretched-out swell, not a sharp swell, what do you mean?"

A: "I mean a long, lazy swell, not close together, steep. No wind chop on top of them."

Q: "No white water at the top of the swell?"

A: "None whatsoever."

From the Coast Guard report, one can point out a seeming discrepancy, based on the record from the Scripps Institute of Oceanography wave rider buoy near the GH sea buoy. That record showed wave heights of 13 feet at 11:25 p.m. Dec. 22, as an average of the preceding 30 minutes, with "probably no diminution in wave energies at the buoy 5 area."

And it was between buoy 3 and buoy 5 that the towline broke, leaving the barge adrift. Lieutenant Emge's USCG report says "the towing hawser parted...somewhere between the winch drum and the tug's fairlead roller."

Captain Freel, on the tug *Janet R*, had towed an empty wood chip barge into Grays Harbor shortly before, and told Lieutenant Emge he had encountered mostly 10-foot swells coming in from the southwest, with an occasional 16-foot breaking swell inside buoy 6, close to the bar. Obviously, there had been some diminution from the sea buoy shoreward.

But the *Ocean Service* and *Nestucca* did not get that far. And the crew's testimony was in agreement that there was no advance warning of the cable break. Captain May, speaking at his deposition hearing relative to shortening the towing cable, said he might have taken in more cable and "may have waited in the vicinity of buoy five for...an outbound ship...to clear the bar. I wouldn't have met him on the bar....I would have picked it up fairly short there and maybe circled or waited for him or whatever it takes."

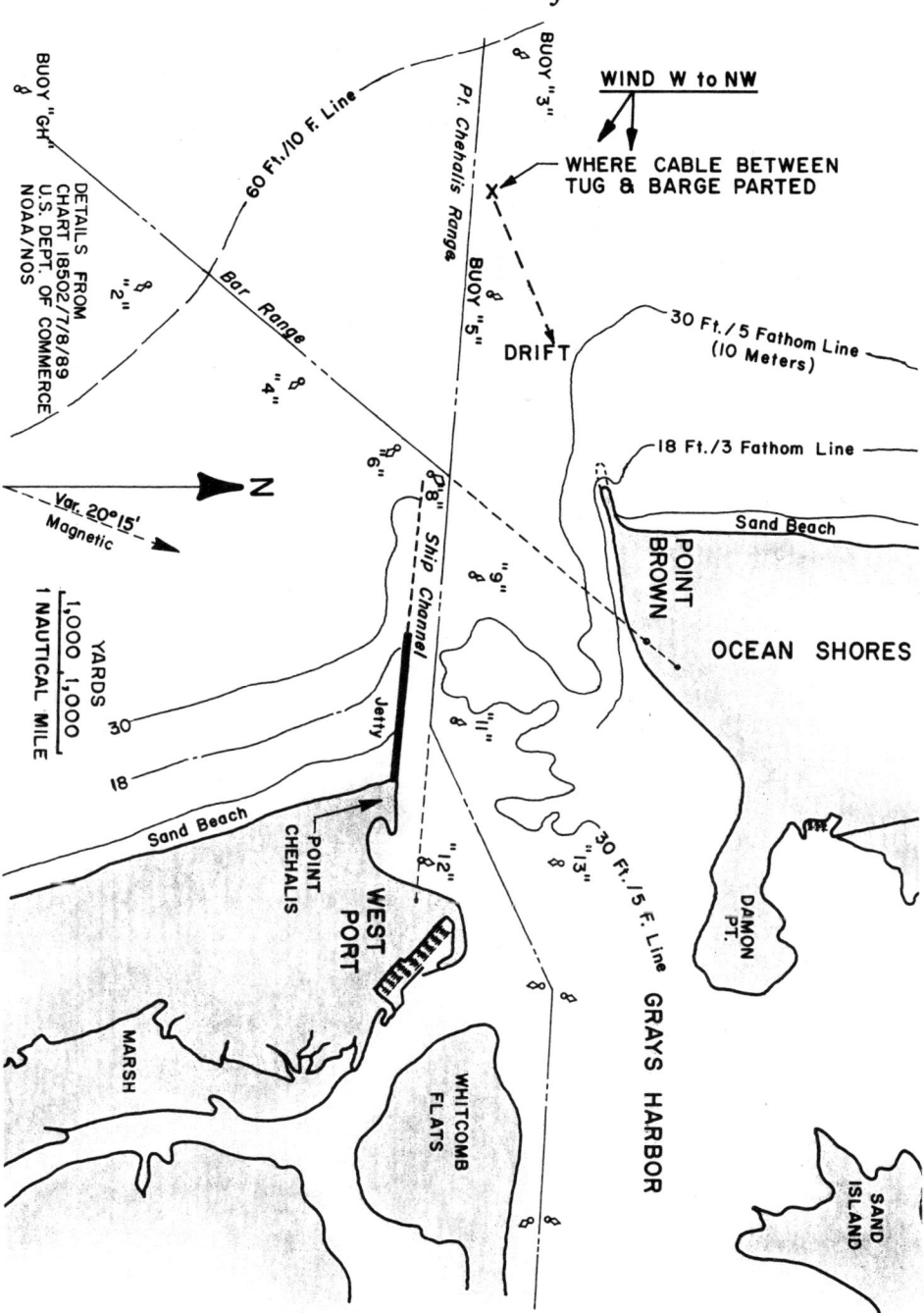

WIND W to NW

WHERE CABLE BETWEEN
TUG & BARGE PARTED

BUOY "3"

Pt. Chehalis Range

BUOY "5"

DRIFT

60 Ft./10 F. Line

Bar Range

"2"

"4"

"6"

"8"

Ship Channel

Jetty

BUOY "GH"

DETAILS FROM
CHART 18502/7/8/89
U.S. DEPT. OF COMMERCE
NOAA/NOS

Var. 20°15'
Magnetic

N

YARDS
1,000   1,000
1 NAUTICAL MILE

30

18

Sand Beach

POINT
CHEHALIS

WEST
PORT

MARSH

"9"

"11"

"2"

"13"

WHITCOMB
FLATS

30 Ft./5 F. Line

30 Ft./5 Fathom Line
(10 Meters)

18 Ft./3 Fathom Line

Sand Beach

POINT
BROWN

OCEAN SHORES

DAMON
PT.

GRAYS HARBOR

SAND
ISLAND

THE ORVILLE HOOK

PATENTED BY:
SAUSE BROTHERS
OCEAN TOWING,
COOS BAY, OREGON

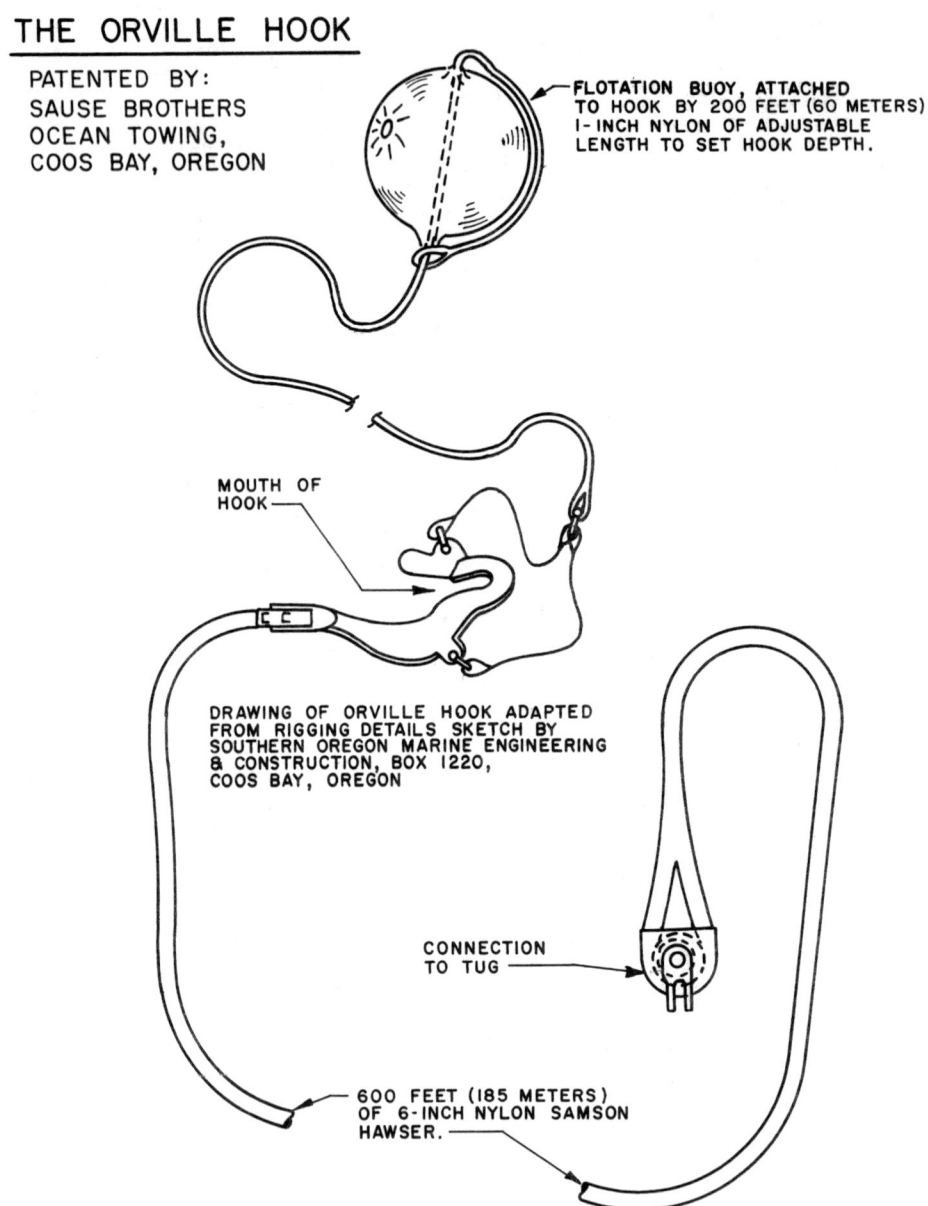

FLOTATION BUOY, ATTACHED TO HOOK BY 200 FEET (60 METERS) 1-INCH NYLON OF ADJUSTABLE LENGTH TO SET HOOK DEPTH.

MOUTH OF HOOK

DRAWING OF ORVILLE HOOK ADAPTED FROM RIGGING DETAILS SKETCH BY SOUTHERN OREGON MARINE ENGINEERING & CONSTRUCTION, BOX 1220, COOS BAY, OREGON

CONNECTION TO TUG

600 FEET (185 METERS) OF 6-INCH NYLON SAMSON HAWSER.

Q: "If there was no ship traffic you would have gone over the bar with the wire at 1,500 feet?"

A: "I may have, I don't know. I would have to get into the bar and look at the conditions."

Q: "In this instance, did you ever get into the bar?"

A: "I never did get in that far."

Q: "And...how would you describe the parting of the wire? What happened?"

A: "The boat kind of took a heave on a swell, I was turned to pull south, probably in a southeasterly direction, and I wanted to get down a little lower on the ranges. But when I was kind of turned into the swell and the boat took a pretty good heavy swell, and at the time that it parted it took a pretty good heave on a swell. When it came back, that's when the wire parted. I could hear it, I heard it by sound."

Q: "What kind of sound did it make?"

A: "A loud popping sound."

The tug's engineer, Jack Wilson, had a different recollection at his deposition. "It was more like a thump.... It was like somebody hitting the side of the boat with a sledgehammer."

About five minutes before the tow cable snapped, Captain May had taken in some 1,000 feet of the tow wire in order to shorten the *Nestucca's* leash for the bar crossing. May also was anticipating the outbound freighter *Southern Star*, which was leaving Grays Harbor with a partial load of logs under the command of harbor pilot Mel Flavel. The *Southern Star*, however, was at that time just below Hoquiam, further from the bar than was the *Ocean Service*, so Captain May probably would have been able to get inside the bar before Captain Flavel with the log ship approached. It didn't happen, because of the break in the towline. Captain Flavel later relayed a radio request for a tug from Hoquiam to take over the empty chip barge *Baranof* from the *Janet R* so Captain Freel could head back out over the bar to help the *Ocean Service*.

Jim Wallace's *Post-Intelligencer* story quoted Captain Flavel's evaluation of events that night: "When the chips were down, they went in and got that barge out. It was very rough and very, very, very dangerous. Once they lost that barge, these people were heroic in picking...it...up." After the towline parted, "Things began to happen pretty fast," as one crewman said at the deposition hearing.

Captain May: "Right after I heard the pop of the tow wire, I went back to see what it was for sure, if the wire broke or the brake had slipped or what. I got back to the back deck and the tow wire is broke, so immediately I went back to the wheelhouse and told the crew, and the chief engineer was sent directly to the engine room to start rigging the Orville hook. The mate and assistant went with me to the back deck to get the nylon out of the box where it was stored, the nylon retrieving line."

Q: "Did you tell the crew in the wheelhouse that you were going to use the Orville hook?"

A: "Yes, I did."

Q: "And who all was there in the wheelhouse when you gave these instructions?"

A: "Jack Wilson, Gary Rickey and Curt Bartley."

Q: "Where was the hook?"

A: "Mounted on a bulkhead in the fiddley above the engine room next to the starboard hatch."

Q: "And how long did you expect it was going to take to deploy the Orville hook?"

A: "Twenty to thirty minutes you should have the Orville hook deployed."

Q: "What...steps would you instruct the crew to perform...to deploy the Orville hook?"

A: "I wouldn't have to...tell them how to rig it, because they know how.... They would rig it."

Q: "...How do you rig the Orville hook?"

A: "It (the Orville Hook) has two wire straps approximately three feet long shackled one to each side of it. From those you shackle--those straps have eyes in them--you shackle a shackle to that with a line going to the buoy, the flotation device, however deep you want to set the hook. Then attach your nylon line to it also with a shackle, it's a rubber shiv with a pin through it."

Q: "When you say a 'nylon line' what line are you referring to?"

A: "Our nylon retrieving line,...300 feet, six-inch nylon."

Q: "It would take 20 to 30 minutes to get that accomplished and get it in the water?"

A: "Right."

Q: "Are you trying to hook the bridle or are you trying to hook the wire?"

A: "You hook the bridle."

Q: "You are actually after the bridle that hangs into the sea behind the fish plate?"

A: "Not necessarily behind the fish plate, depends on how shallow the water you are in...."

Q: "You are trying to hit chain?"

A: "Right, we want the chain, we don't want the wire."

Q: "You are going to be setting the Orville hook to get below the fish plate, that's what you prefer?"

A: "Not really. Bridle chain, whatever is handy, it depends on the situation."

Now take a moment to look at the drawing of the Orville hook (figure 6, page 24), a device invented by a onetime Coast Guardsman and longtime Sause employee, Orville (Bud) Fuller, during the 1960s. Patented by the company, it was installed on all the company tugs for use in emergencies such as faced Captain May and the crew of the *Ocean Service.*

As Fuller recalled for his deposition hearing in November, 1990: "Well, we'd had trouble picking up barges...I'd have to follow a barge around for several days sometimes to catch it and wait for the weather to lay down or one thing or another, and then you'd have to get somebody aboard the barge, so that entailed pretty good weather before you could try something like that. So we were kicking around different ideas and I come up with this one particular one. We had a fellow out here by the name of Bruce Eddy that'd followed a barge around somewhere between here and Honolulu for almost a week, and he couldn't do anything with it, so we were sitting in the office talking to the boss about it and I says, I said I could probably pick that up in about an hour. Or less. Maybe even a half an hour, if I was out there. So he wanted to know why and how, so I drew a picture of it, the hook, and he said, 'By gosh, that might work,' so he had me build one."

Q: "When you say the boss, who are you referring to?"

A: "Curt Sause."

Q: "So you were talking about this with Curt Sause shortly after one of the Sause Brothers captains had spent a week chasing a barge near Hawaii?"

A: "That's right. We'd had trouble before picking them up and several ideas on how to pick one up...like that. Previously we'd lost a chip box (a barge with a large boxed enclosure for hauling chips to be made into chipboard or paper pulp) and it went up the coast oh, overnight from Bandon to Cedar Head Lighthouse dragging all its tow wire, and we had an awful time picking that up. I gave my deed off the tow line (turned over the towing job) to another tug and we got in behind the barge and dropped wire and went around each side of it and picked up the bridle, but we couldn't hold the damn thing after we got it aboard. So I looked at that chain and I thought, by gosh, there's a way to hook that chain. So that's how it came about."

Fuller also commented on insurance wires, the old way of trying to recover a loose barge. An insurance wire is a cable strung from a front corner of a barge back along the side or over the top, with a line and buoy trailing in the water behind. But it did not function well, often getting tangled or otherwise hung up. "...And once you get ahold of them with an insurance wire, why usually you have to jerk the wire off the top of the barge and then, then you've got it by a corner and you can't really pull on it because then you've got, all you've got for cushion is what nylon you have, you know, hooked onto it. And whenever you lose a barge...you've got real tough weather, usually. So you just kind of have to drift along and hurry it along until you get close to a port somewhere and then you have to have somebody else come out and get ahold of the other corner (of the barge) so you can bring it in. Otherwise the damn thing will try to pass you on one side and then it'll come around and pass you on the other side. And then you can't really pull on it, either. So you have to have pretty damn good weather to get it across the bar, you know. So with the hook, why you've got a, you've got the bridle on there for a cushion again. You know, that was the reason why we had the hook."

Now that's from the source, "Bud" Fuller, how he came to construct the salvage hook named for him. One additional observation from Dale Sause was that the Orville hook has never failed in his experience, and that other towing companies call on Sause to recover loose barges when their own efforts fail.

So now we can return to that black night off the Grays Harbor bar, and with Captain May aboard the tug *Ocean Service* peer--at least in imagination--into the dark off to the northeast. There are lights along the beach from homes and businesses on the Ocean Shores Peninsula, but Captain May can also see something more ominous--white breakers between his position and the shore, marking dangerous shallow water not far beyond buoy 5 toward which the barge was drifting.

Q: "Then what happened?"

A: "As we were getting the thing ready, the rate the barge was travelling looked to me as though I wouldn't have time to deploy the Orville hook."

Q: "Why were you concerned about the barge?"

A: "Because it was drifting towards the north jetty...The barge was approaching buoy five. North of buoy five is shallow water, I could look over there and see breakers...constantly in that shallow water. It's a dark night, I can't tell exactly how far it is. The way that barge was traveling it looked like we had no time. If it got into the breakers, it was goodbye barge. We wouldn't have had a chance to retrieve it, any way of retrieving it...The barge never did stop drifting."

Q: "You mean it never stopped drifting prior to the time you recaptured her? Didn't all that tow wire drop to the floor of the ocean there and..."

A: "Correct. It slowed the drift considerably. As we were in here the swells started picking up also. Near buoy 5. The swell started to rise and wind started to pick up at that time in there around buoy five. And with even that much.... We anchor barges all the time on tow wire in flat calm weather in harbors in shallow water, no problem. With a loaded barge and the swell working against it all the time, the tow wire is not going to stop the barge completely, it is going to keep going."

Q: "Do you feel, even with all that tow wire in effect acting like an anchor, that there was a serious risk of the barge...."

A: "Very serious risk of it drifting. I can see breakers...like I say. Would be west of the north jetty. If the barge reached those breakers the whole barge would have been in on the beach."

Q: "What did you decide to do then?"

A: "When it looked to me as though we weren't going to have time to deploy that Orville hook, I told them, 'Hey, guys, get your life jackets, I am going to put you aboard the barge, we are going to have to grab it with a nylon'."

Q: "Who did you tell to get their life jackets on?"

A: "Curt Bartley and Gary Rickey."

Q: "What happened then?"

A: "I gave Gary Rickey a radio, VHF radio, told him what channel I would be on so we could stay in contact. And we had the line right there ready to pass. I went (maneuvered the tug, that is) over to the barge, laid there and watched it for a little bit, and swells were not that bad. I figured, 'Hey, this is going to be a piece of cake, no problem. I am going to pull over there...' The decks were awash naturally because it was a loaded oil barge, there wasn't very much freeboard. So when I put the men aboard... I wanted to put them on as close as possible to the bow shield on the *Nestucca* (see color plate 1, photo section).

"Bow shield is four foot high, kind of runs...around the bow to the port and starboard sides. Pretty good protection from the swells that come aboard. I backed in toward that on the...port quarter of the tug, and they could...stand up on the stern deck which was pretty well matched up...with the barge.

"I pulled in there and a swell kind of banged me into the barge. I told them 'Don't jump, I will get a better position.' I wanted to make sure that the men get aboard safely. I pulled out away from the barge a scat, a little bit, laid there and watched it again and (it) looked good. I backed in there and at that time they jumped. I was back up against the barge. As soon as they were aboard the barge a swell picked me up and set me kind of down on the barge on the port quarter of the stern. I was going full ahead, but it set me back into her anyway. Evidently, as far as I know, that is probably when the barge was holed."

Q: "So you hit the barge twice?"

A: "On both times. You can't get next to a barge in weather like that without...banging it a little bit."

For a different viewpoint, with some variation in details from the way the skipper remembered things, let's turn to the deposition of May's nephew, Curtis Lee Bartley, known as Curt.

Q: "How long did it take you from the time the master said 'Get your life jacket on,' until you got back...with it on?"

A: "Probably two minutes at the most, I was flying."

Q: "Then what happened?"

A: "He backed us into the *Nestucca* so we could jump on."

Q: "How close did he come to the *Nestucca*?"

A: "We hit the *Nestucca* the first time."

Q: "And do you know what part of the tug hit the tow?"

A: "The stern of the tug hit the barge."

Q: "What happened after you hit her?"

A: "I remember getting up from the deck, and we went away from the barge there and came back in and tried it again."

Q: "Getting up from the deck? Were you knocked over by...."

A: "Yes, we hit the barge hard. I got up off the deck, and it seemed like it was maybe a minute and we were ready to try it again."

Q: "At that point is when you jumped on board?"

A: "Yes, we both did at the same time, me and the mate."

Q: "Were you scared?"

A: "No, I was just thinking about doing what I was told."

Q: "Was it much of a jump?"

A: "About ten foot is what I recall."

Q: "Did you make a running jump?"

A: "Yes. It seemed like 10 feet, might not have been that far."

Q: "Did the *Ocean Service* hit the barge any other time than that first effort to put you aboard?"

A: "No, just one time."

Q: "Did you notice any damage?"

A: "When I jumped the second time we came into the barge, when I jumped I seen a hole in the starboard side."

Q: "You saw a hole when you jumped?"

A: "Yes."

Q: "What did you do once you were aboard?"

A: "We went up in front of the (barge's deckhouse)...in behind the splash guard, and was waiting to get a line to the *Ocean Service*."

Q: "What did the captain do then?"

A: "He was maneuvering the boat to get close enough to us, but he had steering problems. I can't remember how we found out, but he might have hollered to us. I think that's what happened, yeah. He hollered to us."

Q: "Was a line ever thrown to you?"

A: "No, we tried to throw a line to them, threw a heaving line."

Q: "Where was that line, aboard the *Nestucca*?"

A: "It was, yes, it was in my hand. I took it from the *Ocean Service*. You always take a heaving line when you board a barge."

Q: "How long a line was that, approximately?"

A: "About a hundred feet."

Q: "How many times did you try?"

A: "Twice, as I recall."

Q: "Was she (the *Ocean Service*) able to get within a hundred feet of the *Nestucca*?"

A: "No. Because that line came about, probably when we threw it, it was about, probably, 35 feet short of landing on the...stern of the boat."

Q: "What happened then?"

A: "Well, the seas were picking up all the time and it got so bad we went around in back of the house on the barge, and we were taking several waves and we finally--the captain hollered at us to get inside the house to stay there until he told us different, that's what we did."

Q: "Do you recall the size of the swells that you were dealing with once you were aboard the *Nestucca*?"

A: "I can't say for sure. They were coming clear over the top of this house...."

Q: "When you say 'over the top of the house' are you talking about spray, green water or how much water?"

A: "We were up to about our necks in it a couple of times. The waves were coming over the top of the house is what I said."

Q: "Are you talking like it's...surf coming in, a solid breaker of surf coming over the bow?"

A: "Yeah, we were hanging on behind the house. I wasn't looking at the waves, I was keeping my head down, hanging on. When the waves were coming over the house we would be up to our necks in water, hanging on."

Q: "How were you communicating with the captain?"

A: "Well, we took a radio aboard the barge with us, but one of the waves hit the mate and shorted out the radio."

Q: "When did that happen?"

A: "Directly after we got on the *Nestucca*."

Q: "After the radio shorted out, how did you communicate?"

A: "The only other time...was when he came around the portside of the *Nestucca* with the boat and hollered and told us to get inside."

Q: "After you went inside the house could you see what the *Ocean Service* did?"

A: "I remember seeing it make two passes around the barge."

Q: "When you made the jump, you testified earlier that you recall seeing a hole in the side of the barge. When did you tell anybody about that?"

A: "I told the mate when we got on the *Nestucca*."

Q: "When you jumped did you see oil coming out of the hole?"

A: "Yes."

Q: "Did anyone ever yell to the captain there was an oil leak out of the barge?"

A: "I can't say for sure. Seems like we did yell to him, but I don't recall him hearing us because he was preoccupied."

Captain May was preoccupied for good reason, so let's go back to his deposition:

Q: "You were intending to get close enough to the barge to receive that heaving line, is that right?"

Capt. May: "Correct. I figured they would hop aboard, I would stand right there, give them the...line and that would be it. Get hold of the barge and be on my way."

Q: "You pulled away from the barge after you put the men on board and then...discovered you had a stuck rudder?"

A: "Correct. We inspected the rudder compartment and there was no visual damage, we weren't taking on water or anything. We suspected that the rudder was jammed, because I...couldn't use it."

Q: "How long was it...before you managed to free the rudders?"

A: "I think I put it in full ahead and held the rudder over and it kind of broke loose, then I had like five degrees left rudder...Five degrees rudder is no rudder, as far as maneuvering."

Q: "Now, how long did it take Wilson after he made the inspection of the rudder compartment, to finish rigging the Orville hook?"

A: "I don't know. I was helping him rig it. I pulled off clear of everything, we rigged it, I set the buoy at the depth I wanted and we threw it overboard.... I went around the barge, came up on the starboard bow...and cut around (the front of it). Jack Wilson was on the spot light so I could see the buoy that was attached to the (Orville) hook. We made a pass across the bridle and snared it."

Q: "Which side of the bridle did you snare?"

A: "The buoy came out a little way and missed the starboard bridle leg and hooked the port bridle leg."

Q: "You were coming around from starboard to portside around the barge...(and) snared the port leg?"

A: "Correct. (And) at that time I had the barge in tow. I immediately maneuvered around in front...and with the limited steering I couldn't hold a course. So I had one engine going astern, one engine pulling, however much power it took to move the barge, stop its drift, gradually towing it out."

Q: "When you talk about setting the length of the buoy (on the Orville hook), what line are you referring to?"

A: "The buoy line that goes from the bridle on the hook to the flotation device."

Q: "This balloon-looking object?"

A: "Correct."

Q: "When you set it, what was your objective?"

A: "Approximately eight to ten feet,...to catch a bridle leg."

Q: "Just so we can understand it, the...bridle hangs straight down, essentially, doesn't it?"

A: "Not in shallow water.... It would have been dragging...at an angle...out away from the barge."

Now, with the connection restored between tug and barge, we return to Bartley and Rickey, soaking wet and probably shivering, taking shelter in the barge house.

Q: "How long did you...spend in the house before you were told...to jump aboard the *Janet R*?"

A: "Probably six hours."

Q: "What appeared to be driving the barge, was it the waves?"

A: "Seemed like the waves to me."

Q: "Did you notice the effect of any current on the barge's drift?"

A: "No, we could see a buoy, we was getting closer all the time. That's why I knew we were drifting toward the lights."

Q: "Could you...feel when...the *Nestucca* was reconnected to the *Ocean Service*?"

A: "We had a good idea. We could tell by the noise that we were being towed. After the *Ocean Service* made its second pass, we seen it going around and we started hearing a lot of noise from the chain and stuff, the bridle chain off the *Nestucca*. And that's when we figured we were being towed."

Q: "Can you tell how soon after you went into the house that this second pass occurred and you felt like the *Ocean Service* had grabbed the bridle or the tow wire?"

A: "Probably 30 minutes, I guess. Twenty to thirty minutes."

Q: "Did the buoy stop getting closer?"

A: "Yeah, with watching that buoy close, we would go way past it and come back out from it, towards the end there. We kind of figured we were going the other direction when we came out away from it."

Q: "Was it just once you went past the buoy and came back out?"

A: "Yes, as I remember."

Q: "So you think that when you went past it you were adrift, when you came back out would be under tow again?"

A: "Yes, that's when we figured we were okay."

One other matter of interest in young Bartley's testimony was the following:

Q: "Was your uncle (Captain May) at all panicked or (did he) appear to be panicked at all?"

A: "No, he knew exactly what he wanted to do. He always does."

Another piece of testimony that helps one to understand the situation and succession of events that night is this exchange with Mr. Wilson, the tug's chief engineer:

Q: "Did you see any hole in the barge?"

A: "No, I didn't."

Q: "Was it quite dark at the time?"

A: "Well, it was 11:30 at night, but we have floodlights on, floodlights on the stern of the boat. You could see the barge all right. I don't think it was really dark, no." (Note: The time was close to midnight, as the towline had parted about 11:30 p.m.)

Q: "Do you recall approximately how far the mate and the assistant had to jump to make it from the stern of the tug to the barge?"

A: "I think they jumped when the barge and the boat made contact, so there was no horizontal distance. Maybe vertically two feet they jumped down from the boat onto the barge."

Q: "How much time elapsed between the time the men were put aboard and the time you abandoned trying to get the heaving line and went back to...rig the Orville hook?"

A: "Well, for him to get in front of the barge and then make two passes and then discover that the steering was stuck, it was probably five, ten minutes."

Q: "How much time then did it take to get the Orville hook...ready to deploy?"

A: "Five minutes, maybe."

Q: "How long did it take...to actually snare the tow wire (in fact the bridle chain) using the Orville hook?"

A: "After we had it in the water, he just went around the barge and hooked it, so ten minutes maybe."

Q: "Now, after you retrieved the tow...using the Orville hook, what did you do?"

A: "The skipper started pulling it away from the beach. And when the barge was out of danger, we hooked it into the (end of the broken) tow wire where I had put an eye in it and let the tow wire out so that the nylon retrieving line wouldn't chafe on the stern, and we towed it down towards--out to sea, I think."

Q: "How about visibility, do you recall seeing any surf or breakers?"

A: "Yes. It wasn't foggy. We had good visibility. I think there was an occasional rain squall, but there was good visibility. You could see the lights on the beach and (all) that."

Q: "Did the weather worsen that evening? Did it get better? Did it stay about the same?"

A: "When we first lost the barge, the weather was good. And then as we were trying to recover the barge, it worsened. And then it started getting better again after that."

Q: "When was the weather at its worst?"

A: "I think it was worst...probably about the time that we tried to recover with the Orville hook."

Q: "And by worst, could you describe the conditions for us?"

A: "The wind picked up. There was more seas and rain squalls. But it still was only six or eight foot swell. But it was worse than it was prior to that and after that."

Q: "Can you estimate the height of the seas?"

A: "Two to three feet on top of the eight to six foot swells."

Q: "And the force of the wind and direction?"

A: "Oh, 15-20 (knots). But the wind comes and goes with the rain squalls. When the rain comes, the wind blows harder. And when the rain goes away, the wind dies off. And so you can have a patch of bad weather move through and it will just be like a local disturbance."

Gary Rickey, the mate, recalled the swells as having been a bit higher, "eight to ten foot," as he and Bartley stood on the tug's stern waiting to jump to the deck of the *Nestucca*. He recalled the jump as being "about six feet, maybe," in between the estimates of Bartley and Wilson.

As to the radio he carried, Rickey said at his deposition hearing, "It shorted out almost the instant I got on the barge, within a few minutes. We had a wave come over the bow where Curt and I were standing and it went over our heads."

Relative to whether they tried to shout to Captain May from the barge to inform him it was leaking oil, Rickey said his memory was vague. "You see, we got hit by a wave right after we got on the barge; and you kind of shut down when that happens to you. It's real cold."

That observation by Mr. Rickey recalls the fact that neither he nor Barkley wore a survival suit for the leap to the barge, even though the *Ocean Service* and all Sause tugs carry them as standard equipment. Time was the essence in the skipper's mind, and donning survival suits would have been time enough to deploy the Orville hook. Captain May estimated there was not time enough before the barge would be in the breakers off the end of the north jetty.

The crew's depositions and testimony at the Portland trial in U.S. District Court on liability for the spill did not focus at length on events subsequent to recovery of the barge. But information given by Captain Freel of Astoria, skipper of the tug *Janet R*, helps to complete the story, along with some details from Captain May's deposition.

The *Janet R* had towed the empty chip barge *Baranof* into Grays Harbor shortly before the *Ocean Service* broke its towing cable, and waited off Westport until a Foss tug could come out from Hoquiam to take over the barge. Then Freel headed out to sea again to assist the *Ocean Service*.

It was well past midnight by then, and the tide was ebbing. Freel told Coast Guard investigators later that he encountered eighteen to twenty-foot swells, with every other wave breaking. The drama at sea was being monitored ashore by radio, both by Coast Guard Group Grays Harbor and by Mrs. Josephine (Jo) Dyas, a veteran of more than 10 years handling communication to and from vessels in the Grays Harbor vicinity. A contractor for NOAA (the National Oceanic and Atmospheric Administration), she provided weather information to mariners.

As the *Janet R* pulled in behind the *Nestucca*, which was creeping seaward behind the *Ocean Service* a half mile south of buoy five, Captain Freel radioed to Captain May at 2:04 a.m.:

"Oh, my God, Chuck, you're leaking oil!"

That was the first anyone, except for the two men on the barge, knew of an oil spill. Freel also reported considerable oil already on the water around the barge.

What had happened was that the left rudder of the *Ocean Service*, which is twin screw with a rudder astern of each prop, had ripped a six-foot gash in the side of the *Nestucca's* number one tank. That tank had a 6,000 barrel capacity, and was loaded with 5,800 barrels (243,600 gallons) of bunker fuel oil. Although the hole was near the top of the tank, seawater poured in as the oil poured out, and the heavier water went to the bottom of the tank, lifting the remaining oil so that more could flow out. By the time the hole was plugged 23 hours later at sea, less than 20,000 gallons of bunker fuel remained in the tank.

With knowledge of the spill, Captain May's difficulties multiplied. He had hoped to pass off the tow to Freel and the *Janet R*, but things were suddenly complicated. May needed to get in touch with his headquarters in Coos Bay, but was unable to. He and Freel discussed by radio the possibility of turning over the tow to the *Janet R*, and Freel told May that the *Nestucca* was a large barge for the smaller tug to handle; and if he took it, he would plan to tow it in to Grays Harbor on the next flood tide, about noon Dec. 23. Freel also did not want to chance the changeover until daylight, because of the combination of darkness and rough sea conditions.

Captain May said at his deposition hearing that he would not have passed the tow over to the *Janet R* "until somebody from the company (Sause) advised me to do that."

May also asserted, "I wouldn't have taken that barge in there under any circumstances on any tide with the setup I had. That barge would have to be passed through somebody else, I wouldn't attempt it."

He was referring, of course, to the fact that the tug's only connection to the barge was the 600 feet of nylon clamped onto the port bridle chain so that the *Nestucca* was being towed at an angle at very slow speed, and his steering control on the *Ocean Service* was limited almost entirely to adjusting the speeds of the tug's twin engines. Not only that, but about 1,500 feet of the broken towing cable was dragging on the relatively shallow bottom, with only 40 to 50 feet of water beneath the vessels.

In addition to these considerations, word was radioed to the *Ocean Service* from Group Grays Harbor that Washington State Department of Ecology officials did not want the leaking barge towed into Grays Harbor. Their concern was potential damage to oyster culture areas and the National Wildlife Refuge west of Hoquiam, a major migratory shorebird stopover. The Coast Guard, heaving a collective sigh of relief at the narrow escape from having a barge with 3 million gallons of fuel break up on the North Jetty, directed that the *Nestucca* be towed out to sea.

"We headed down the coast with it," Captain Freel of the *Janet R* recalled, "and it was pouring oil out, and there was too much swell to do anything."

One other event relative to his arrival on the scene stuck in Freel's memory of that night. "When we went up to the barge, I've never seen anything like that in my life. These guys were in that bow compartment, and it was completely under water I would say 75 percent of the time. Those guys threw the door open, came steaming out of...I mean, they were in sheer panic to have us come and get 'em off. And I said 'No, no. You guys just stay in there and we'll wait until daylight'."

It was about 7:20 a.m., before sea conditions moderated some- what and Freel was able to take the *Janet R* alongside the *Nestucca* to transfer Barkley and Rickey to the tug.

"Lucky thing the tankermen had left some clothes in there (in the barge's cabin) and kept them dry. There was two or three changes of clothes, and they went through all of them. Then when we got them back on our boat, we made 'em shower, because they were damn near blue. Stuck 'em in the shower to warm up. Put coveralls on them, then we cleaned up their clothes, dried them out. Later in the day we managed to get up alongside Charles and get them back on their boat."

During that morning, a Coast Guard helicopter dropped a damage control kit onto the after deck of the *Janet R*; but it was not until 11 o'clock that night, 23 hours after the *Nestucca* had been gashed by the tug's port rudder, that the weather moderated enough for repairs to be made.

"It was the middle of the night," recalled Captain Freel. "The swell had come down (to 8-10 feet, says the Coast Guard report), and I asked Charles to put her over, get it so that it was kind of giving a lee on that one side. I asked our guys, 'Hey, does everyone feel safe about going out there?'

"So I stayed on the boat, running the boat, put two guys on the barge, Monte McCleary and Robert Eddy. Bobby, he was...it was...you know, there was still enough slop there...Bobby could reach the hole over the side if he lay down on his belly on the barge, and Monte held onto Bobby with a rope. Well, Bobby was under water most of the time driving wedges.

"We tried using oil spill diapers first to...take up some of the slack, but then Bobby said he was getting...I was talking to Monte on the radio between the boat (and the barge). Bert Little was in the wheelhouse with me, holding a searchlight. Bobby said the diapers were causing more problems than they were doing any good, so he just ended up patching it with wood. The Coast Guard gave us a whole duffel bag full of shims. With 2-by-4 wedges, you can imagine how long Bobby was in the water.

(The Coast Guard reported the gash in the *Nestucca* was 6 to 7 feet long and up to 6 inches wide. It was about 18 inches below deck level.)

"Bobby didn't have a survival suit, all he had was a life jacket and rain gear. Monte, holding the rope, had a Mustang survival suit, one of those Mustang work suits. But the difference in height and weight between the two guys, Bobby couldn't get in Monte's suit. I don't think Monte was gonna give it up anyway," Freel recalled.

Summing up his opinion of events involving the *Ocean Service* and *Nestucca* that night of December 23 near Grays Harbor buoy 5, Captain Freel said, "As far as I'm concerned, between the guys who were left on the *Ocean Service*, when they got their line on there and then got their rudder screwed up, and they were still trying to keep things going, still had ahold of that barge, I mean...you can't believe what was going on on the bar that night, as far as weather. We were...in the middle of the ebb (tide), and we were right in the middleground there on the bar, and it was damn nasty. There's no two ways about it, those guys deserve something for what they did.... The bottom line is, he kept the barge off that North Jetty."

Recalling later events as they headed southwesterly to a point off the Columbia River, Captain Freel said the *Nestucca*, *Ocean Service* and *Janet R* were even further offshore than had been reported previously by others: "About 36 miles, it seems like. When we got the hole patched, we were 35 or 36 miles off, and they told us once it was patched we could come on in."

That is 58 kilometers, well beyond the continental shelf into deeper water where the north-flowing winter current is stronger and more swift. NOAA's HazMat experts said, however, that the southwest wind pushed the oil back toward shore, where it was carried north by the current coming out of the Columbia River. In mid-January of 1989, as oil ebbed and flowed with the tides along the west coast of Vancouver Island, word was passed belatedly to the Canadians that the leaking barge had been as far as 40 kilometers offshore.

"One of the things that seemed real ironic to me about this whole thing," observed Captain Freel, "was that Charlie towed that thing from Grays Harbor to clear down here off the coast, and then we got the hole patched. So they let us come back in to the beach, and we were gonna bring it across the (Columbia River) bar, and we got to the sea buoy and they wouldn't let him bring it across. They made us take it away from him and start towing it in, then the *Salishan* (another Sause tug) came, and she got ahold of the other corner (of the barge), too. We started in from just outside the sea buoy, and they had an oil tanker coming down, one of the big tankers coming out of the Columbia, and the captain of the port called me and told me I couldn't go any further than the sea buoy. I said 'Whaddya mean?' and he said 'Well, this oil tanker's coming down, and you haven't got clearance to enter.'

"Well, I got mad," continued Captain Freel, "and took it all the way in to Number 2 (3.1 miles outside the bar), and I called him back and said 'Look here, we only got two more hours of flood (tide), I'm going to get this barge inside. I can take it up to Sand Island, stay out of everybody's way; but I'm not gonna sit out here another tide.' Well, we went on in, and when we got to the bar, they came out with a helicopter and Coast Guard boats, and they wouldn't let us proceed any further until they checked the patch. There must have been 12 guys...went over and looked at the hole and got on the barge and looked at the patch. By that time the swell had dropped way down.

"We went on then, and I think we were right around (buoy) 12 (3 1/2 miles further, directly south of Cape Disappointment and Ilwaco, Washington), and Charlie got made up in the notch behind (to push). We got up to right around 21, between 21 and Desdemona Sands Light (another 3 miles upriver but still 7 miles short of Astoria, Oregon), after that tanker got by. Then we turned around, and we sucked in their insurance line (actually the Orville hook's nylon hawser), piled it onto the deck of the barge, then took off, and we went home for Christmas. That was the end of the story for us, they cancelled our charter right there.

"It was Christmas Eve. And the worst thing was, the following year, Crowley (another towing company) lost an oil barge on the bar here, and we had to stay with it until the 24th. Two years in a row I was out there tending oil barges."

So the *Salishan* and *Ocean Service* took the *Nestucca* upriver, first to Astoria and then on to the shipyard at Portland, where both tug and barge were repaired before going back in service (see color plate 3, photo section).

British Petroleum (BP), owner of the *Nestucca's* cargo, advised the Coast Guard that approximately 227,304 U.S. gallons of oil were lost, plus or minus 100 barrels, 4,200 gallons. The maximum loss, then, would have been 231,504 gallons, which was 7.87 percent of the total 2,940,000 gallons (70,000 barrels) the barge carried. Minimum loss would have been 223,104 gallons.

Mention of the barge's capacity brings up a confusion factor that arose in the first days after the spill. Marine parlance deals with oil by barrels, but landlubbers use gallons. An early release to news reporters mistakenly employed the barge's 70,000 barrel capacity as *gallons* rather than barrels, and that figure was used by news media in some reports as the quantity of oil spilled. The actual loss in the spill was not determined until the barge arrived in the Columbia River, probably after it reached Portland.

Having discussed background aspects and the actual events of the *Nestucca* spill, we next consider how and why the oil spread. This includes public response, or lack of it through unconcern or ignorance, contrasting with anger or the positive "hand me a tool to help." Then we'll progress to real life and the factual impact of oil on our coastal beaches and wildlife.

How the Spill Spread

Public awareness and understanding of efforts to track the oil spilled from the barge *Nestucca* were both frustrated by inaccurate or incomplete information.

The Washington State Department of Ecology's duty officer, Bill Young, was contacted by the U.S. Coast Guard from its Portland office at 12:30 a.m. December 23 and advised that the barge was leaking "black tar crude" (actually bunker-C fuel oil) on the Grays Harbor bar (3-1).

In fact, the leaking barge was more than two miles west of the bar, probably some five or six miles out to sea at the time, creeping along at perhaps two knots or less on an emergency hawser behind its tug, the *Ocean Service*.

Even if the barge could have been hauled into Westport on the next slack low tide at 6:19 a.m., before daylight, a patch kit would have had to be delivered by helicopter from the Astoria Coast Guard station. This was done later that morning to the *Janet R* at sea. Perhaps it could have been done earlier.

Information relative to how far offshore the leaking barge was eventually towed by the *Ocean Service* was probably inaccurate and wasn't even reported to all concerned parties. According to Capt. Freel of the *Janet R*, the three vessels were 36 miles (58 kilometers) offshore when the patching job was completed and they were given Coast Guard permission to head in. The State DOE report said the distance off the coast was 28 miles. Whatever distance is correct, the course on which the *Ocean Service*, *Nestucca* and *Janet R* were moving was southwesterly, into the wind. Although beyond the edge of the continental shelf by the time the *Janet R's* crewmen were able to stop the leak in the barge, 23 hours after it had been holed, the spilled oil was mostly pushed back toward shore by the southwesterly wind. That information is according to the Hazardous Materials Response Division oilspill experts of the U.S. National Oceanic & Atmospheric Administration (NOAA).

Another aspect of tracking the spill was the combination of bad winter weather conditions and the difficulty of spotting oil on the storm-tossed ocean. There were reports that most of it floated below the surface of the water, leaving it extremely difficult to see from the air. Specific gravity of the *Nestucca* oil was measured at 0.99 in a later laboratory test. Bunker-C oil, also termed residual fuel oil number 6, has an official specific gravity of 0.95 (approx.) at 20 degrees centigrade, 68 degrees Fahrenheit.

"'We thought we had it whipped,' said Doug Zimmer, a State Department of Wildlife spokesman. 'The beaches at Ocean Shores were cleaned, the oil was offshore and there was a feeling we were in the fourth quarter of a football game, leading 30-0. I think of it as a malevolent monster that's out there. It moves, adapts and changes'." (3-2).

A Dec. 26 report that proved to be in error was from the U.S. Coast Guard to Canada. It stated that the 1.5-by-3-kilometer slick spotted off the Queets River (120 kilometers from the international border) was estimated to be moving at a rate of six kilometers a day and "would not reach Canadian waters before breaking up." (3-3)

At best, that report was misleading, because quantities of *Nestucca* oil came ashore along both Olympic National Park and Vancouver Island, beginning Dec. 30 and 31 (see figure 2, page 8). The oil drifted as far northwest as Moore Island, off the Canadian mainland in Hecate Strait beyond Queen Charlotte Sound some 700 kilometers (about 430 miles) from Grays Harbor. It was found there January 24, 1989, which was 33 days after the accident. That represents a northwestward drift averaging 13 miles or 21 kilometers per day, more than three times faster than the Coast Guard's estimate (see figure 2, page 8).

Relative to tracking and predicting the movement of the heavy oil, Coast Guard Canada's *Nestucca* oil spill report said:

"The response team could not predict with certainty where, when and in what quantities the spilled oil would come ashore on Vancouver Island...or whether it would proceed further north to the mainland or the Queen Charlotte Islands. There was also a great deal of controversy surrounding the cause of the re-oiling of beaches.

"Initial predictions of the anticipated movement of the spilled oil were provided by the U.S. Coast Guard and were derived by the National Oceanic and Atmospheric Administration using a mathematical model. Attempts to track the oil by visual sighting, infrared and ultraviolet sensors and synthetic aperture radar were not very successful. Based on these efforts, it was concluded by the USCG that the oil had dissipated and it would likely have little impact on Canadian shorelines except for the stranding of occasional small tar balls.

"Institute of Ocean scientists (in Canada) concluded that, based on (early) information that a small spill had occurred off Grays Harbor, the oil would move with the upper layer of the ocean and would be carried...toward Canada by the northward-flowing currents (of)... the coastal flow in winter months.

"The presence of oil...was confirmed when it washed ashore at Carmanah and on Long Beach. Institute of Ocean scientists used this information coupled with their understanding of the seasonal and wind-induced currents to predict...movement... further northward...to the tip of Vancouver Island and beyond into Hecate Strait.... At the time it was not known that the barge had been towed offshore into the relatively fast-moving currents of the continental shelf break. Also the barge had leaked nearly six times as much oil as had been reported earlier.

"...It appears...that the oil was spread over a very large area...as a result of the barge having been towed over (beyond) the continental shelf while...leaking and by the dispersion effects of...turbulence in the ocean currents. Further, the oil had broken up into pieces...obscured by...wave action and...too small to detect.

"The impact...was that oil stranded on the shoreline in a sporadic fashion. Areas of heavy contamination could be surrounded by areas with little or no oiling.... There was also the...uncertainty that there might be more oil at sea waiting to impact...already cleaned beaches."

One of the suggestions made in connection with the spread of the *Nestucca* oil was that of Coast Guard Canada for employment of satellite-tracked buoys with sub-surface drogues such as are used for measuring ocean currents. Air deployable buoys also are available, according to CCG, that can be tracked by aircraft.

Coast Guard Canada's description of how *Nestucca* oil spread and then came ashore was, however, not acceptable to members of the NOAA "HazMat" team, as they are known. NOAA's leading expert on oil spills and their spread in the marine environment is Dr. Jerry Galt, at the Sand Point national headquarters of the Hazardous Materials Response Division at Sand Point on Seattle's Lake Washington shore. Galt said the keys to the oil affecting both Olympic National Park and the west shore of Vancouver Island were more complicated and less publicized than was indicated by the Canadian Coast Guard report.

A genial red-bearded veteran of over 15 years in oil spill prediction and research, Galt explained that understanding how and why the *Nestucca* oil spread requires knowledge of several complicating factors, including ocean currents, wind direction, the action of oil slicks on water, and the little-recognized aspect of 'convergence zones.' These are outflows from rivers and bodies of water such as Puget Sound that meet and intermingle with ocean currents flowing along the coast.

He said the oil spilled from the *Nestucca* was pushed back toward shore by the southwesterly wind and was not caught in the swifter ocean currents beyond the continental shelf. The oil was moved northward along the coast by the current flowing out of the Columbia River, but about the end of December was pushed offshore by a strong east wind blowing out Juan de Fuca Strait, caused by a high pressure area that moved down from Canada. The oil spill by that time had largely broken up into tarballs, Galt explained, which is typical of bunker fuel, so that it was not possible to track it from the air. Spread over a broad expanse of storm-swept ocean with both windchop and swells, tarballs haven't the typical oil sheen and are invisible from the air.

"There really isn't any mystery about this at all, except...that strong east wind pushed it offshore, just when everybody was glad to see it gone," observed Galt. One might point out in passing that Christmas was over and the new year at hand, so those involved might well say of the spill, "Goodbye."

But the *Nestucca* oil, even though widely scattered offshore by the east wind, was reconcentrated after resumption of the normal winter southwest winds pushed it toward Vancouver Island. As Dr. Galt explained, the reason the widespread tarballs regathered was that the northward-flowing coastal current meets the flow of brackish (mixed fresh and salt) water coming out Juan de Fuca Strait from Puget Sound, and the combined streams bend northwestward along the coast of the 288-mile-long island (467.5 km). That meeting of the waters is termed by oceanographers a convergence.

The result was that the stretched and spread-out swarm of tarballs from the December 22-23 oil spill gathered back into a concentration, and under the push of the onshore prevailing wind came ashore on Long Beach, the 15-mile stretch of Pacific Rim National Park between Ucluelet and Tofino, B.C. The remainder of the *Nestucca* oil continued moving northwest along the coast, and more oiling occurred as onshore winds pushed tarballs onto remote island beaches.

"What we were not quite ready for," said Galt, "was the ferocity with which this turned around. We thought it would be strung out more. As it got into the stronger (near-shore) current, it would get stretched more. And what happened was the winds got almost perpendicular to the coast (of Vancouver Island) and were pushing it (the oil) right against the beaches. So instead of blowing it along the coast, with a light spatter all the way, it targeted pretty hard the area of Long Beach and Barkley Sound.

"There's no such thing as a small oil spill, if it's your beach." Some of the tarballs, Galt said, will coalesce into oily pancakes of variable size, especially on sandy beaches. Subsequent tides often will bury such stranded oil during winter accretion. Buried oil can add so much sand to its composition that it will not later float away because of its increased weight.

As to the oiling of the Olympic National Park beaches, Galt said that the area from the Norwegian Memorial north to Yellowbanks, the beaches west of Lake Ozette, received the heaviest concentration. This occurred just prior to the easterly wind event that pushed the spill away from the coast, so that the further shore on toward the Makah Reservation was spared.

Dr. Jerry Galt's explanation of the NOAA HazMat division is helpful to understanding their manner of working. Someone from the division, Dave Kennedy in the case of the *Nestucca* spill, is assigned as technical staff to the U.S. Coast Guard's on-scene commander (OSC) for each oil spill to which the CG responds. The staff at the NOAA Sand Point headquarters serves as backup, tracking the spill with their computer programs, often using information gathered from the air.

"Years ago," Galt explained, "we set ourselves up as a group which does research and response. Our fundamental charge is to respond during spills to minimize environmental damage and to support the Coast Guard on technical questions. The third aspect of our job is to do research and, if possible, to do better next time. The research (we do) is appropriate to the kinds of problems we ran onto in the last spill, and so over the years we've developed techniques and procedures.

"We work on a number of different research programs, such as how to deal with the various types of radar and sensing devices used in aerial surveys. We also have active training programs for our own people, and now we've taken people from some of the states that have asked us. For 15 years we've conducted training programs off Santa Barbara, California, where they have natural oil seeps.

"We've developed digital geographical positioning machines (GPS), global navigational gear. We have a hand-held electronic box with a keyboard, so you can fly out and you put an overlay on the keyboard and while you're typing it records on a floppy disk, and when you come back you stuff that into your computer and it draws out where you've been and what you have seen. We do a lot of computer models."

In the course of discussing convergences, Galt recalled the *Exxon Valdez* crude oil spill of March 1989 in Alaska. "We tracked the oil down through the (Prince William) Sound, out into the Gulf of Alaska and all the way to Gore Point (near the south end of Kenai Peninsula),...between Prince William Sound and Cook Inlet. At that point we lost the spill as a coherent slick. It was scattered patches. From there to the Barren Islands,...the biggest piece of oil reported back to me was a guy said 'I saw a patch the size of my dining room table.' We had numerous reports of small globs. So here we are again with it spread out, but we know it's not over. We said, 'Let's think about this.' It turns out that the fresh water coming out of Cook Inlet is just about like what comes out of the (Juan de Fuca) Strait here, and that same kind of convergence zone is...there in Cook Inlet. So all of a sudden one morning we get a report back that out at Katmai there's a slick 25 miles long and 25 meters wide. That was the convergence zone that picked up all those tarballs and brought them back together. This was almost 300 kilometers...from the source, and the spill had been gone for 120 kilometers in terms of anybody being able to see anything except spatter on beaches. And when it came ashore there were a bunch of places that got pretty heavily hit. Windy Bay, Tonsina, these were places that got a lot of oil because of an onshore storm.

"The amazing thing to me," continued Galt, "was that 80 percent of the bird kill occurred right in that apex. So despite the fact that you think the spill is mostly gone, spread way out, all over the place, it can reappear, causing serious problems, and it's a death trap for birds. The birds...killed were the ones that live on the surface, and at night they sleep, and they are pulled...into the same convergence as the oil. Knowing where the currents are is a minor part of the problem. You have to know where the convergences are.

"Let me make another point here, relative to putting out buoys," continued the NOAA expert. "There have been a lot of experiments, but buoys don't work in tracking oil spills. What happens is, the surface drifts because of wind and currents, and anything floating in it is moving along also. Oil spills, on the other hand, are slick, not rough, like the difference between corduroy and cotton. Your eye picks it up in a flash. It takes very little oil to do that. You know the bit about putting oil on troubled waters. All these little waves are absorbed by the oil. If you go to the beach and are hit by a wave, it's no surprise that waves have energy. And what happens is, as these waves push into oil they shove it along, so the oil is actually moving faster than the water."

At this juncture, I asked, "You mean the oil actually picks up energy from the water, it's imparted from the waves to the oil?"

"Exactly," said Galt. "We've done this experiment a number of times. We can demonstrate it using dye in the water with time-lapse photography...The oil is going to be (moving) at a different speed than the water. A buoy (dropped into an oil spill) is not going to move the same way (as the oil). Now they've done things like putting a little sail on the buoy and a blade sticking down. And the Canadians market something (that) looks like a little frisbee, with a flange or turned-up edge, and their argument is the waves reflecting off this edge give the differential motion that it needs. But the problem is that the oil spill at this stage is a fluid on a fluid, but very soon the oil will break up into streaks and streamers that have a very different wave profile, and the waves don't radiate through it in the same way, so that the push is different. And the third stage of the oil spill as it breaks up has a wave profile nothing like the previous stages. Each stage is different. The wave-oil interaction does not remain constant. So that, even though in theory one can design a buoy to work quite well in one situation, there is none that can work in all of the different stages of an oil spill. Although this sounds like a good idea, our experience with them has been pretty disappointing," concluded Galt.

Returning to the subject of the *Nestucca* spill spreading northward to Vancouver Island, Galt displayed a series of computer printouts demonstrating the day-to-day location of the oil. "We can see, on the 27th (of December) it starts to move offshore...then it relapses, and here's Juan de Fuca (undersea) Canyon, and by the first (of January) we had reports from all over of hits on the shore...I wouldn't deny that some of (it) might have cut across (from the Juan de Fuca Canyon deflection) under the influence of the wind (and landed at Carmanah Point on Vancouver Island). On the third (of January) it starts coming ashore on Long Beach. My view is there's nothing magic about it. Bunker oils have a very long 'threat distance' because... they...will continue to spread...unless there is some mechanism that brings them back together. They...spatter any beach along the coast. Beaches on the East coast are a lot worse than on the West coast, and beaches in Alaska are definitely not as bad as down here or as bad as those in Southern California or the Gulf Coast. Stuff spread out over a hundred square miles of ocean is trivial, but when it comes ashore it looks bad...You never beach an oilspill with a current, because currents run past a shoreline, not onto it. You must have a wind...to carry it ashore... We did not expect in any sense that this would not come ashore up (in Canada) someplace. But we...expected it would be spread out so far that it would be indistinguishable from the (normal) tarball population. ...Any day, you're going to get a few tarballs. What (we expected) would happen is, you'd get a few more."

Galt conceded that the 150-200 oil spills to which the HazMat team responds annually "probably (makes us) a bit jaded in this regard (referring to tarballs spattering beaches)."

Inspecting the printouts of the spill, I noticed they showed it as starting from Grays Harbor, running southwestward to a point offshore, then turning southeast to the mouth of the Columbia River. The latter part of that, of course, did not happen, since the leak in the barge was plugged before the Coast Guard would permit it to turn back in toward the coast. But Galt said HazMat does not go back after the event and correct such an error that had been based on misinformation.

A much stronger negative opinion is voiced by Galt relative to public reactions to oil spills. "As soon as you have an oil spill, the next thing you have is a media spill, then it becomes a bureaucrat spill because all the people who handle that suddenly get replaced by their next-ups. Then as soon as it becomes a big deal you get a politicians' spill, and what happens then is, you're at the spill and you can't get a helicopter because they're touring the brass. And all of that slows...the response. And then usually it's a big money spill, and that confuses the issue a great deal, with everybody trying to figure (to their advantage)...and ultimately you get into a big lawyer spill, and that is to me a real disaster. The problem is, lawyers come in and become advocates. Lawyers aren't interested in the truth. They're interested in the subset of the truth that supports their case. And the way this comes out is that information...on the spill is declared 'litigation sensitive' and withheld. What that means is the cleanup people, the guys who are really trying to fix it, are dealing with only half a deck. And I could care less whether...anybody...gets the(ir) pants sued off...on account of the spill. That's not an issue with me. My concern is, how are you supposed to do the right thing...looking at half the data? And...environmental issues associated with spills are so terribly complex that either side...is very reluctant to take it to court.... Spill after spill does not actually go to...trial. The various sides...leak facts, and they show either that there's no problem, everything's okay, or alternatively that everything is going to hell, and life as we know it is going to end... The lawyers have become big-time spin doctors, and your hired scientist is not free to talk about half the problem. That has tremendous influence on public perception, which has a tremendous influence on driving the response. And once again, the people...trying to do something about it (the spill) are being yo-yo'd around because of perceptions...carefully planted for another reason. Again, it's a money spill, nobody's going to declare it over as long as there's money to be made.... And nobody's going to admit that they're--a term somebody used the other day that I laughed at--that they're 'spillionaires'."

Asked about Thom Davis' observation that there haven't been enough spills the last couple of years to keep the experienced Global Diving & Salvage crew intact, Galt observed: "This is a problem...in the technical community, in the response community, in the press corps and everywhere. Spills are...rare enough...that...for the vast majority of people who deal with one, it's the first they've seen.

"We went through an...argument years ago in NOAA when we set up this organization, about whether we should be local or national. Parts of NOAA...are local, like the weather service. The Seattle office, for example, forecasts weather to the Columbia River. South of there is somebody else's weather. There was a pretty good argument that they're on duty 24 hours a day, they do weather, they should do oil spills. The question came down to 'how many weather offices are there around the country? The training bill is going to be outlandish. Chances of those guys doing two spills, particularly two big spills, is almost zip.' So this is a national technical support group here; but we have one-person offices in Boston, New York, Norfolk, Miami, New Orleans, Long Beach, Seattle and Anchorage."

With the ocean's intricacies of currents and mind, we've learned how the spill spread, and the factors that brought heavy oil back ashore in such concentrations at great distances from its origin. Now we move on to the lives, animal and human, impacted by the oil.

4

Cleaning the Oiled Beaches

Response to the *Nestucca* spill started the morning of December 23, 1988, less than 12 hours after the barge had a hole ripped in its side by the port rudder of its towing vessel.

The Washington Conservation Corps put 43 of its workers on the Grays Harbor north beach before noon that day, and while one crew started bagging clumps of oil, the second crew began walking the beach to pick up both live and dead birds that had begun to wash ashore. Kirk Thomas, On-Scene Director for the WCC, said that fortunately some of the young people on the crew had recently learned bird cleaning techniques, and a few of them helped teach volunteers arriving to assist with the oiled birds. (4-1)

Thom Davis, one of the co-owners of Global Diving & Salvage in Seattle, recalled the circumstance. "I had done a spill for Sause Brothers probably two years before...up at Cherry Point (refinery near Bellingham, a Mobil operation at that time, but now part of the BP system). They had spilled a bunch of oil on the deck, and we went in and cleaned it up. That's how I got to know Dick Lauer, but it...surprised me when he called.

"It was kind of funny because it was the Friday before Christmas, and we have a day off at our shop. What we have done for several years is...take all the company trucks down to Northwest Harvest, which is a food bank, and help make deliveries for them. It's kind of a fun thing. So here I am driving down the 99 viaduct (Alaskan Way along the Seattle waterfront) with a load of turkeys, and I get a call (on the car phone) from Dick Lauer. He explained...that they had gotten the barge back, and that I should get down to Grays Harbor and take a look as soon as possible. That was about 11:30 in the morning.

"So I called Queen City Helicopters, friend of mine, told him I would be at Boeing Field shortly. I went and unloaded the truck, then back to Boeing Field, got in the helicopter and we flew down to Grays Harbor. It was rough going down there, because it was so windy. Real rough day, and we were going to try and make it back by dark. Couldn't see the barge, it was already out to sea. I could see little traces of oil, not a whole lot, it wasn't anything too exciting. Saw a little oil right there between the breakwater and Sand Island, then a little bit from the breakwater up to the hotels (at Ocean Shores, next to the outside ocean beach). I was on the cellular phone in the 'helo'

(Davis pronounces it "heel'-oh," meaning of course helicopter) getting the guys organized back here. That was a tough time, because we had a spill that was going to take us at least over the Christmas weekend, and it was essentially the same thing that had happened to us on several occasions before. A lot of guys have kids, and they know they're going to be tied up over Christmas, so it was not easy. But we got down there, brought all our gear out.

"On a beach spill like that, there's not much gear involved, because it's pretty much all manual labor. There's no mechanical device to clean up oil on the beach. We take 4-wheel drive vehicles, trailers with ATVs (all-terrain vehicles), pitchforks and shovels. Plastic bags are a key item. Before *Nestucca* we probably kept no more than 500 or 1,000 bags in stock. Ever since we keep about 25,000 on hand, because by the time we began to run short we emptied every store at Ocean Shores, and of course they are not the quality of bag we want. We must have used 50 to 60,000 bags. Well, we have our own bag made under our label now, a 6-mil bag, real heavy duty, 50 to a case.

"The thing that was a new experience for us--I never had understood this before--but Grays Harbor's in the Portland zone of the Coast Guard, up as far as Queets. Working with the Portland Coast Guard was something we hadn't done before. We had worked with the Seattle Coast Guard and were on a first-name basis, understand each other. The first guy...from the Coast Guard was not real easy to get along with, but he was gone the next day. A little lieutenant, nervous about what kind of a spill this was, didn't know who we were. He threatened federalization on the spot, wanted 300 people on the beach Christmas morning, which of course was an impossibility.

The second Coast Guard guy we had was Lt. Guy Theriault, a big tall fellow, good hiker, real easy to get along with, and the third was Greg Yaroch (pronounced Yair'oh) from Seattle. And the NOAA representative was Dave Kennedy, head of their spill response team. That was a neat thing about *Nestucca*, because we had Kennedy and Yaroch along with Lew Kittle and Jim Oberlander (from the State Department of Ecology). That team had worked together several times before."

Initial beach oiling was in the area just north of the Grays Harbor entrance, plus Sand and Goose Islands inside the harbor. Over the ensuing week, the oil spread further north with each high tide, but progressively lighter oiling on the ocean beaches helped confirm a hope that the drifting oil was dissipating and would be no further problem.

Heaviest oil impact on the Washington coast, according to the USCG report, was in the vicinity of the Norwegian Memorial and Yellow Banks beaches as far as Sand Point and beyond in the Cape Alava area. South-facing beaches took the brunt of the *Nestucca* oil.

Seattle Post-Intelligencer reporter Lisa Schnellinger wrote of her January visit to the Olympic National Park beaches in a news feature story (4-2):

"Strange sounds fill the beaches, an undercurrent below the crashing of oddly dark waves. Shovels chop piles of kelp, rakes scratch the sand, rubber gloves shoosh the thick blobs of oil into little rolls like pastries and plop them into plastic bags.   And helicopter blades saw the muted misty air and the heavy green silence of the nearby forest.

"Such sounds have not been heard often before on Washington's stretch of wild beach. The Olympic National Park's 57-mile coastline is unmarked by development.... People dressed in yellow rubber suits are cleaning up oil at Sand Point, near Ozette, and Third Beach near LaPush....

"There is something else different about the beach music. There are no bird cries. There are fewer birds. The birds that remain at Sand Point and Third Beach have mostly been sorted already into plastic bags. One for live birds, one for dead ones. Next to them are the other bags, also sorted. One for oil blobs, one for plastic trash, one for wood and other burnables. The bags get picked up by helicopter in loads of 1,000 pounds or so.... They go to a dumpster for tabulating. Then the sandy oil goes to be made into asphalt, and the rest is incinerated.

"Crews have been working at Third Beach since...Sunday, but the marshy saltwater beach still smells like a gasoline station. It's slow work, and the standards are higher here because the stakes are higher, explain the owners of Global Diving and Salvage, the cleanup company. Since the pristine beach is so sensitive, workers will take up blobs as small as quarters, says Tim Beaver, one of the owners supervising at Sand Point. National Park and state officials will inspect to see if the cleanup is good enough, and if not the crews will return.

"'The oil companies spent a lot of money training us,' says Thom Davis, another owner, supervising at Third Beach. 'They really know what they're doing, and they want us to do a good job.' Davis is proud of his company, and he thinks that in a short time the damage at Third Beach won't be visible....

"Kevin MacCartney, National Park ranger at Ozette, is concerned about the sea otters. There are only 100 in the state, and they congregate just north of Ozette at Cape Alava, where oil has yet to be picked up. 'The otters depend on their fur for insulation, and they can't clean the oil off themselves,' MacCartney says. 'They could die of hypothermia.'

"...When he looks at the sheen of oil, MacCartney also sees ammunition. This may be what the Park Service needs in its fight to stop oil drilling and mining along the fragile coastline. 'We're afraid of what offshore drilling could do to this,' he says, looking around at the rugged sea stacks and tall trees that frame the water.

"The damage isn't immediately obvious. But on the wild beaches, oil is easy to pick out--it gleams and shines in the dull gray light. Gobs of oil--like sand dollars--cling where they were sprayed and drooled by the waves. Oil bleeds from the sand and drips from the driftwood. Birds are still washing up in clumps like dirt, oiled beyond recognition. Workers call them 'dippers' because they look like they've been dipped in oil.

"While they're cleaning up the oil and picking up birds, the workers pick up the rest of the litter. There are motor oil cans and old shoes and little shampoo bottles from the cruise ships. There is even a 55-gallon metal drum that washed up at Third Beach.... A sign at the trail head warns visitors to stay away from the metal drums because they may contain toxic materials.

"But mostly the cleanup work comes in gradually smaller bits--clumps that come up with the incoming tide. Bits of grass, bits of animals, bits of human trash. In a pile of seaweed coated with oil, two wings are barely distinguishable. A grebe, his eyes gone, guts picked cleanly by a predator, lies cradled in the curve of half an auto tire. One scoter, his eyes glaring out from his blackened body, runs away from a human and dives into the surf. A murre with limbs askew sprawls in one solid slick, like a black plastic toy. Above it all, a bald eagle flies silently back and forth. Back and forth."

In a report from the Quinault Reservation, John Dodge of *The Olympian* (4-3), told of tribal concern about bald eagles seen on the beach eating oil-soaked seabirds. Normally, the eagles are up at Lake Quinault this time of year, eating the carcasses of spawned-out sockeye salmon, Dodge was told by Larry Workman, tribal forester.

"A walk on the beaches south of Point Grenville on the Indian reservation...offered tell-tale signs of the state's second largest oil spill. Oil oozed from the tidal pools tucked next to the sea stacks that jut skyward on the beach. Dime to quarter-sized pieces of algae and wood coated with oil dotted the beach, grim reminders of an unwanted visitor. In some areas near Point Grenville, the oil washed ashore in large enough sludgy globs to allow cleanup crews to retrieve it.

"Dead birds were wedged in the debris left by high tide, awaiting removal by the volunteer crews. 'I found 11 dead birds Wednesday and eight more Thursday,' volunteer rescue worker Sandi Phares of Stanwood said.

"At least 8,000 oiled birds have washed ashore from Oregon to Vancouver Island, including 4,500 dead ones," *The Olympian's* story continued. "Fortunately, few other forms of wildlife appear (to have been) affected by the oil, officials say. 'There is no evidence of problems with the razor clams,' said Doug Simons, a shellfish biologist with the State Department of Fisheries. He said the high and low tides have been high enough to deposit the oil above the clam beds. In addition, the bunker oil is so thick it is unlikely to penetrate through the sand to where the clams live. 'The oil spill occurred in prime Dungeness crab fishing grounds, but crab fishermen have yet to find oiled crabs in their crab pots,' Fisheries shellfish biologist Steve Barry said. The crabs have escaped...because the oil doesn't appear to be sinking to the ocean floor,...but...crab pot buoys...have been fouled by the oil.

"A small population of sea otters off the north Washington coast has escaped the oil so far, State Department of Wildlife spokeswoman Janet O'Mara said. Flights over the otter habitat...revealed some oil, but no oiled otters,...which numbered 136 in 1987.... 'The oil will collect in the kelp beds and the otters forage (there),' she said.

"About two-thirds of the 60 miles of beach in the Olympic National Park had patches of oil, said the park's chief ranger, Chuck Janda.... 'You will see it if you're hiking on the park beaches next year, but probably not five years from now,' Janda said. 'This is an historic first we could...do without.... It is a good example of the park's vulnerability to outside forces'."

Janda since has retired after a long career with the National Park Service. In February, 1993, Janda recalled in an interview at his Port Angeles apartment how "We...(had gone) along with all the media reports, and they--the Coast Guard--were trying to play it down. There was nothing to fear, the currents would take care of it, it was well offshore....

"But the first inkling we had was just a bomb. There was this feeling I got in the pit of my stomach when Lew Kittle (of the state DOE) called me...New Year's Eve. That was the first indication that oil was showing up on our beaches. Kittle called about 6:30 in the evening and requested permission to bring equipment onto Third Beach." (Some 3 miles, 5 kilometers, south of LaPush.)

"Oh, it was spotty. Ten beaches were significantly oiled, but (it) was observed in at least minor concentrations along nearly every mile of the 57-mile coastal boundary. It came ashore in scattered concentrations as opposed to continuous sheets.... Most...was deposited at or above mean high tide line.

"On sand beaches, it formed intact globules or mixed with woody or plastic debris. Pebble, cobble and rocky beaches were the most visually impacted, the oil coating the rocks over extensive areas. In some areas of heavy drift accumulation, hundreds of logs were splattered to greater or lesser extent.

"During subsequent tides, much of the oil became mixed with beach substrate, though still retaining the cohesive characteristic. Oil globules were buried as much as two feet on sand beaches and three feet in pebble and cobble. On boulder and bedrock beaches, some of the oil has...weathered or has been abraded (so)...that only dark stains remain.

"Maybe the place to start explaining how it affected Olympic National Park is with the land. At the time of the spill, ONP had exclusive jurisdiction of the coastal strip from mean high tide shoreward, meaning that the park has exclusive authority to enforce all federal laws. No other agency has that authority. With a national park what you have is the land set aside for the enjoyment of the public in perpetuity. It is in fact a sacrosanct area that is not to be violated in any way, shape or form."

Webster: "Are some of the national park ownerships checkerboard like so much of state ownership?"

Janda: "Well, checkerboard, I think, would be too regular. Patchwork would be more accurate for national parks. There are very few parks that you could say are pure in the sense that parkland ownership is total. I worked 14 years in Glacier Bay National Monument in Alaska that is almost that way.

"The complicating fact on the coastal lands is that they are statutory wilderness. On top of that, it's a world heritage site. In spite of all these protections, you are totally out in left field when it comes to an oil spill. And the narrow coastal strip included in the park you could not call a complete ecosystem, so our management strategy has been as best we can to try to preserve this primitive coastline.

"We have only a few private developments along this entire park coastal strip--Ozette, a little at LaPush and then down at Kalaloch. Anyhow, at the time (1988-89) we were going through negotiations with the state (of Washington) to...acquire title to the tidelands from high tide seaward to lowest low tide. Then there was legislation that gave the park title to the offshore island refuges. Now they are officially part of the park, although administered by U.S. Fish and Wildlife. That jurisdictional thing gets complicated and it's probably a side issue. We were concerned with the beaches, and the entire critical intertidal area, all the way out."

Webster: "Just in passing, I've seen no mention of any cleanup being done on the offshore rocks and islands."

Janda: "No, never. They did a very extensive survey from the air of these offshore rocks and islands, to determine if there was any significant...oil pollution, and concluded that yes, there was, but that it would be virtually impossible to get out there and clean it up. I suspect it was spotty, and that the surf conditions would largely take care of it.

"In any case, access to this coast is very limited, almost no roads. There are just a multitude of informal trails (starting from logging roads in areas east of the park boundary) and our rangers use them on patrol. We have some very heavy restrictions on use of aircraft even for rescue purposes. It is not something we take lightly. If it takes a couple of days longer to walk in and walk out, that's okay. We accept the land on its own terms. We avoid using (rental) helicopters for beach access except in emergencies, and prefer foot patrols because of the detrimental effects of aircraft on the environment of the park.

"Many of the agencies we dealt with during that spill had no concept of what a national park is or was intended to be. Of course, you would expect that. So we had a lot of raised eyebrows when we would at various times impose limitations on use of equipment or methods, but we never got a whole lot of static on it.

"As chief ranger, I was a division chief, reporting to the park superintendent, who has overall authority and responsibility. I had authority and responsibility for all aspects of resource management and park visitor protection, including law enforcement, search and rescue, fire management and any other disasters such as an oil spill.

"From the time this first started, I was Park Service--I guess you would say--representative within the oil spill organization. Yet at no time were we considered members of this oil spill response team."

Webster: "You were consultants?"

Janda: "It was even hard to put us in that category. We were a resource trustee, sort of like a homeowner who reports a fire, and the firemen show up and say 'We'll take over from here. You stay out in the street or go see your sister or something. We'll take care of this fire.' It's almost that you have to entrust other entities with your interests. These folks were saying 'Just sit back, have another cup of coffee, we'll take care of it.' And of course we're not willing to do that."

Webster: "You realized from the start, I suppose, that their standards of cleanup were different from what you require."

Janda: "Oh, absolutely. Our standard is no oil. It requires no more explanation than that. No oil. In any case, I mobilized my ranger force. We have people stationed along the coast. Among our ranger staff of 25, about 20 were actually on outside ranger duty, and six were along the coast, two each in three districts and subdistricts: Ozette, Quillayute or Rialto (Mora) and Kalaloch. All we could do at the time was to work with the contractor (Global), the responsible party (Sause), the Coast Guard, State Department of Ecology and State Department of Wildlife, and the U.S. Fish and Wildlife, all these, and try and figure out how we were going to do this.

"But at the outset--this sort of set the tone for the cleanup, I'm sorry to say--on Monday, the second of January, 1989, when I showed up, that's when we had our little encounter with the Coast Guard. We had indications of a massive oil spill, there wasn't any question about it anymore, and the question I had was 'What are you guys going to do about it?'

"Well, at the time they reported they had 12 or 13 people from the contractor out on the beaches cleaning up. And I said, 'My God! You can't be serious. We have an entire coastline affected. You're telling me 13 people are going to clean up this mess? Let's be realistic.' They got quite defensive. It would have been good to have had a tape recorder there, it got a bit hot and heavy. I demanded that this oil be cleaned up in reasonably fast time to...standards which we would be happy to describe, and explained that this was not just any piece of real estate, that it is a national park, and so on and so forth.

"At one point I was not getting any kind of favorable response, and I said 'Well, within the next two days I will have a Park Service response team on the beaches cleaning up, and that's it.' I was ready at that time to walk out."

Webster: "You would have authority to go out and hire the necessary people?"

Janda: "Yes. You always are walking on a few eggshells there, because you're deficit spending. There is no money. We have contingency planning money for handling fires, law enforcement, but not for this sort of disaster. So we would have had to use--as it turned out we did use--operating funds for our involvement. We were later reimbursed. They found money back in Washington, D.C., Park Service money.

"But at that meeting January 2, the (Coast Guard) lieutenant told me I could not do this, that what I was proposing was illegal. He cited various and sundry things, among which was--I believe it's an executive order--requiring that the federal government not compete with private enterprise. In other words, if the Park Service had gone in there we would have been competing directly with the contractor, Global Diving & Salvage, taking money out of their pocket. Well, I don't think Global was all that concerned about it. Thom Davis of Global...(is) a wonderfully efficient person. He did a good job on *Nestucca*. Nevertheless, we wanted the job done, so eventually he doubled the number of people, and we cooperated with him, provided logistical support, cargo nets, helicopters."

Webster: "But that's still not a big crew."

Janda: "It's not. Their reasoning was that if they had larger numbers of people it would be too difficult to coordinate, it would be unsafe, people might be stranded by the tide, caught in the surf and so forth. The contractor was totally unprepared for remote areas. These guys had typically done harbor spills, with direct access for heavy equipment and so on."

Webster: "You were his training ground for this sort of situation."

Janda: "Exactly. Of course Park Service personnel under our Incident Command System are equipped and trained for that. We could have had an extensive organization working, our own personnel from the park maintenance division. I could probably have got somewhere close to 70 or 80 people out on the beaches the next morning, and provided logistical support for them. But I did not have the monetary resources to do it.

"Now there's another little kicker,...the 311-K fund. It's money the Coast Guard has available to use on oil spills. The... fund at that time had been pretty well depleted, but there was still...money in it..., The Coast Guard did not want to open this up, and we requested it several times,...because...Sause was willing to continue...the cleanup effort, which they did.... But the fear was that if the Park Service got in on this, Sause would pull out. Under maritime law, they were only responsible for expending up to the value of the barge and its cargo, and that of course had been blown in the first few days.

"Well, anyway, between the lieutenant (Lt. Guy Theriault, OSC for the Portland CG District) and myself, we had some pretty good arguments. I was not going to deny that the Coast Guard had authority in the situation; but at the same time, the resources at stake were so great that we could not afford to take a back seat on this. Basically they were going to let the responsible party and the contractor do it, at his own pace, to his own standard at his own expense. And I'd go for one of those things, but not three of them.

"Seems to me that on three different occasions that evening...he had to...call the port captain at Portland and get advice. Each time he would excuse himself and go back to the telephone.

Webster: "Now this is interesting,...because Portland handled the original spill; but at some point (the Grays Harbor-Jefferson County line, just south of LaPush and the Quillayute River) cleanup supervision...(went) to the Seattle Coast Guard District."

Janda: "Well, after January 5 there was Lt. j.g. Duane Smith from the Seattle CG District's operations office and Lt. Greg Yaroch, the Seattle OSC representative, both pretty good to work with. Well, they were all nice people. But what was so frustrating to me was that the Coast Guard was so afraid the responsible party was going to pull up stakes and leave if we looked at him cross-eyed. They went out of their way to treat him with the utmost respect. Whatever Sause Brothers wanted to do, that was okay. And we could not accept that. We wanted things done faster, we wanted them done better, we wanted them done with more efficiency, with less impact on the landscape, a whole raft of concerns....

"But...you see the relationship between park administration policy and the Coast Guard policy, and where the contractor stands and the responsible party. Very inflexible and complex relationships. And here we are as resource trustee of what is invaluable land, this irreplaceable park property.

(An additional party was Lew Kittle, OSC for the Washington State Department of Ecology. Kittle capsulized the three cleanup policies as "to the extent feasible" for the Coast Guard, "to the condition prior" for DOE and "no oil" for the park.)

"We could have been somewhat flexible in the beginning," Janda continued, "but as it stood, we had to play by the rules. The thing dragged on.... It was a real education to...visit the command center (at LaPush), see all these officials sort of standing around, the Coast Guardsmen sitting by their telephones and computers and such. I never figured out what the Coast Guard actually did. Really. They were there. And a couple of times they conducted cursory searches for missing contractor personnel. But aside from that, as near as I could tell, they were giving the responsible party no direction as to 'this is what you should do' or anything like that. I never heard anybody give the responsible party an order, directive or even a strong urging. If the responsible party felt like doing it, then it probably would get done. But it got into a policy of cajoling, flattering, suggesting.... I didn't go for that. But because of the level where I operated, mostly with Thom

Davis and some of his people, why, things got done.

"It was always...the responsible party was not willing to spend any more money than he absolutely had to, and the Coast Guard was afraid to urge him to do so because he'd pull out, then they'd have to open up the 311-K funds, which they didn't want to do. In the meantime, I was accruing these costs out of my operating fund, and I didn't have it. Something like $6,000, and I was being tight," said Janda.

In fact, 311-K money *was* used to clean oil from Protection Island and the area around Dungeness Spit and the mouth of Discovery Bay, in Juan de Fuca Strait 80-85 miles (140 km) east of Cape Flattery, after Sause declined financial responsibility.

The On-Scene Coordinator's Report from the Marine Safety Office for Puget Sound to the Commander, Thirteenth Coast Guard District, was dated October 26, 1989, but a copy was not released until a 1992 Freedom of Information request was made on behalf of this author from the office of U.S. Rep. Jolene Unsoeld, member of Congress from Olympia. Official author of the report was Capt. H.H. Dudley, pre-designated OSC, who had assigned Yaroch as OSC, assisted by Smith. This report, while barely referring to the "how clean is clean" problems, did provide information on Protection Island cleanup.

Oil was found in the Protection Island area Jan. 9, 1989, and later confirmed by laboratory analysis to be from *Nestucca*. As a result of Sause refusal to take liability (presumably because NOAA said it could not possibly be from a spill 80 miles west and headed north, with no eastbound current to carry it), a separate pollution case was established Jan. 11, and the Coast Guard hired Foss Environmental for the cleanup (4-4). Beaches of the affected U.S. Fish and Wildlife Preserves on Protection Island and Dungeness Spit, and around the mouth of Discovery Bay were "lightly" oiled with patties seldom exceeding 12 inches diameter. A total of eighteen 55-gallon drums of oiled debris were collected in the area at a cost for cleanup and disposal of $2,835 with a recommendation to recover that amount from Sause. The work was completed January 18.

Kittle's recollections of the cleanup corroborated in many ways Janda's account, but with a differing viewpoint. As the On-Scene Coordinator for the State of Washington, he worked closely with both the Coast Guard OSC (either Theriault or Yaroch, sometimes Smith) and with Dick Lauer, the On-Scene representative from Sause. The trio came to be known as the "three amigos." It was they who made the decisions, at the conclusion of the daily "1700 meeting," on work to be done the following day. Those sessions, each afternoon at 5 o'clock, involved all persons concerned with the spill and cleanup, including Davis of Global Diving & Salvage, the contractor; Janda or his representative, Ranger Howard Yanish; county officials, tribal fisheries biologists, U.S. Fish & Wildlife district biologist Don Kane or his designee, and often other people.

Reports of oil on any beach would be assigned at the 1700 meeting to a response crew to check out the following day and, if needed, to do the cleanup. That procedure avoided the four-day bureaucratic delays experienced in British Columbia.

A key to keeping control of the cleanup, said Kittle, was that the state held title to the intertidal zone along the beach strip of Olympic National Park. Title since has been transferred to the park. When asked what he was doing supervising cleanup in ONP, Kittle replied "Look out there, that tideland belongs to the state. It has oil on it, so we're cleaning it."

"New" oil that arrived on the beaches later than *Nestucca's* also was recalled by Kittle, who agreed with the Canadian reports that it was obviously different. (Laboratory testing reportedly was inconclusive.) He heard rumors that it was Alaskan crude from a passing tanker. The new oil was cleaned up along with the old, and crews were also instructed to remove plastic and other unnatural debris found on the beaches. This added considerably to the total volume of material.

Kittle was impressed with Janda, and in fact believed that the chief ranger had personally hiked the entire length of the Olympic Park ocean beaches. Queried about this, Janda laughingly said no, he had not been able to take the time to cover the 57 miles himself. What Janda did have was reports from individual rangers who took personal inspection hikes of shorter segments repeatedly during the cleanup. Janda said he was dropped off by helicopter for a few relatively short inspection hikes. "Someday I'd like to hike that entire coastal strip of the park," Janda said.

Lew also said he observed Janda's frustration and discomfort at not being at the helm of the beach cleanup. Kittle kept agents of the Department of Ecology on the beach constantly during the cleanup, depending on Diane Harvester and other on-site observers.

Relative to keeping Sause on the job as financier of the beach cleanup, Kittle agreed that the tug company was "iffy" about staying with it and concerned with limiting the cost. He told Lauer that if Sause pulled out, he would use state money to continue the cleanup, with the implication that Sause would be billed later for the expense. Kittle learned only afterward that the state fund he had counted on could not have been used for such a purpose. "But it worked," he observed with a grin.

Only small quantities of oil were removed from the Quinault Tribal Reservation, a 24-mile stretch of coast between Moclips and Queets, just south of the ONP boundary. Kittle explained that much of the reservation beach is narrow and fairly steep, only 50 feet wide at low tide and composed of gravel. What oil stranded on south-facing beaches often was taken off by ensuing tides and carried elsewhere. Immediately behind the beaches are 200-foot bluffs which make access difficult, so reservation cleanup was limited mostly to the areas on either side of the streams that drain the upland and empty into the ocean: the Quinault, Raft and Queets-Clearwater rivers.

Jim Oberlander, spill response team leader for Ecology and Kittle's assistant on the *Nestucca* cleanup, recalls knee-deep oil on a Quinault Reservation beach. This was partly corroborated by Larry Workman, tribal forester. "At one of our 5 o'clock meetings at Ocean Shores," said Oberlander, "tribal people came walking in wet and cold and said they'd found major concentrations of oil. Kittle and I went out there next day with Guy Tyrrell of the Coast Guard, and we had to rappel down this bank, and it was raining hard. I went up the beach with Larry Workman, and we found tar patties up to our knees. But when we returned the next day, they had all gone, refloated and, I assume, moved north."

Workman said the concentration was on New Year's Day at Hogback/Boulder Point, just south of Raft River (4-5). "Nearly a mile of beach was covered with patties between 12 inches to 4 feet across. The smell of gasoline filled the air. But the following day, the beach was as clean as ever, although there were some oil splashes on the rocks and driftwood encrusted with black goo. Where the oil went became a mystery. It simply disappeared, perhaps part of that heading north.

"On January 3," the QNR account continued, "the Crowley Environmental Services...crew hired by Sause Brothers began the first major cleanup (on Quinault land). That day they removed four tons of oily debris from just north of Point Grenville. ...At Whale Creek over the next week (they) removed nearly 1,000 bags totaling nearly 20 tons from along our beach."

Sand beaches are relatively simple to clean with shovels for scooping, rakes to gather small globules and plastic bags in which to put the oil. Large patties often can be rolled and then sliced into manageable sections. "Pebble, cobble and rocky beaches were the most visually impacted, the oil coating the rocks over extensive areas," according to Chief Ranger Janda's written summary for Olympic National Park.

"In some areas of heavy drift accumulation, hundreds of logs were splattered to greater or lesser extent. During subsequent tides, much of the oil became mixed with beach substrata though still retaining the cohesive characteristic. Oil globules were buried as much as two feet on sand beaches and three feet (one meter) in pebble and cobble."

Another form of oil burial occurred on Sand Island in Grays Harbor's North Bay, where wind covered the pancaked oil with dry sand. That low-lying island, built of silt deposits carried into the harbor by such streams as the Chehalis and Humptulips, is less than 10 feet elevation at high tide, a sand shoal held together by the roots of marsh vegetation. Buried oil is not visible from the air, so cleaning required extensive walking. Sometimes the wind would expose an oil patch also, according to Kittle's OSC Report (4-5). Where a buried layer of oil was at an eroding edge, pieces would break off, leaving a fresh face of tar for cleanup workers to spot.

Sand Island was considered important for cleanup because it is a summer nesting area for Caspian terns as well as gulls. It is also a haul-out area for resident harbor seals. Oberlander recalled seeing many seals in the area, but said they seemed able to avoid being oiled. "Perhaps they smelled it," he suggested.

From Grays Harbor to Moclips, some oiled logs were burned right on the beach, while others were hauled to designated sites for burning. Inside Olympic National Park, oiled logs were lifted by helicopter and carried to burn sites outside the boundary. Large logs had to be cut into manageable chunks for pickup. The DOE report lists more than 584 tons of oiled waste disposed of at the LeMay landfill east of Aberdeen and about 45,000 cubic yards of oiled logs burned.

In addition, about 90 cubic yards of oiled pompoms were added to hog fuel at the ITT Rayonier Pulp Mill, Port Angeles, and burned. Thereby hangs a tale of an intriguing development in oil spill recovery. Thom Davis of Global Diving & Salvage was given credit by Janda for developing the use of plastic pompoms strung on lines to pick up oil on high-energy rocky beaches that are next to impossible to clean manually. The pompoms are waste from the manufacture of insulated wire. Quite similar to the pompoms used by cheerleaders, but heavier, they are tied at intervals of a couple of feet along a nylon line that is then stretched over a section of beach and anchored in place. Sometimes a network of lines can be laid out in a grid covering a considerable area. When the tide is high, oil washed from boulders and cobble is deposited on the pompoms.

Heavy with oil, the pompom lines were separated at low tide into 60-to-80-foot manageable sections, lifted from the beach by helicopter and carried to a plastic-lined dumpster at the nearest disposal site. Davis explained that sometimes pompom lines partially buried by the tide on a sandy beach could be pulled free by a helicopter. The choppers were limited to 1,000 pounds or less in the loads they could carry from the beach to the dumpsite. With a full load of fuel their lifting limit was about 600 pounds, and it was the responsibility of the beach 'slinger' (loadmaster) to know how heavy to fill the cargo nets. That required close coordination with the pilot. Each beach crew had five cargo nets to work with, one being emptied at the dumpster, one in the air and three on the beach being filled.

"You gotta get the most out of that chopper, it's running you 500 bucks an hour," said Davis.

Another innovation was suggested by Janda to clean off oil that had been thrown by the surf high onto dry rocks above tideline. He persuaded Davis to have a portable steam cleaner flown to the beach. "He didn't want to do it, but I insisted; and it worked like a charm, it really did, on rocks where oil adhered and caked so that it could not be wiped or scraped off," recalled Janda.

"Did Janda tell you about the worker who left his initials high on a rock after working with the steam cleaner?" asked Kittle. No, Janda hadn't mentioned it. "When we spotted that, the man had to climb back up and get his initials off."

Davis also introduced another cleaning method with boulders that had been oiled underneath. Workers rolled them down the beach and left them exposed to the surf on the rising tide. In combination with pompoms, that was effective.

These techniques, Janda observed, were "largely due to the fact that Davis and the Global crew became more accustomed to the landscape, working with it, along with the rapport we struck up. Howard Yanish, district ranger for the coastal strip of the park, was a big factor in that.

"The contractor here was quite aware of...buried oil, and they were digging in the sand for oil patties. They took oiled logs out. No burning on the beaches here.

"This was not intertidal area with barnacles or mussels or other life. In the intertidal zone you would occasionally spot a sheen, but there were no globs in that area, just up around high tide line and beyond in the drift stuff and on dry rocks.

"We felt," continued Janda, "that if you were going to have an oil spill, this was about the best you could get, because you had virtually no toxics in the heavy stuff left after they (the refinery) had removed the goodies.

"My strongest recollection is that there is no really good way of dealing with an oil spill. I was very critical of other agencies at the time," Janda continued, "but when you come right down to it, every agency has its own agenda, and there is no central authority or philosophy to put it all together. They each have their own turf, and you better not tread on it. They're going to react. And yet, at the same time, you look at the significance of this land belonging to the people of the United States. And the people involved in this oil spill in many cases were not treating it in that fashion. They were thinking of it in terms of the job they had to do, and viewing it in a very narrow...way."

Discussing the fact that no physical evidence was obtained of the oil damaging beach marine life, Janda explained that, "About two weeks after the spill there was a massive die-off because of cold weather, a very hard freeze. This doesn't happen very often on the outer coast. But a lot of marine organisms died, and we never found any traces of oil in these. The other thing is that there is very little baseline information. More recently we have studies being done, but we didn't know the populations of sea urchins, anemones, mussels and so forth."

Because of the lack of physical evidence, ONP did not take part in the court action against Sause. "The only thing we could have done, and we considered it long and hard," said Janda, "was to go on 'spec,' that is the land had been desecrated by the oil, and there is a monetary value to that. We would be going out on a limb, because it had never been done before. The final decision was, after some months,...that there was no way I could convince the government (to take) this course of action. The chances of winning this argument on the esoteric values of wilderness in a national park were questionable, although we'd make headlines if we won. But I could not get the commitment, never ran across an attorney in our office who was really gung ho, eloquent on behalf of wilderness parkland, and we needed someone like that.

"We came to the conclusion, well, the salvage work has been done the best we can. The beaches are clean, they're done to the highest standard possible. We're satisfied, we got our $6,000 back and we will benefit by some of the research projects. But I don't feel ethically right trying to get as much money as we can," concluded the retired chief ranger.

Thom Davis' recollections add interesting details to the story of the *Nestucca* cleanup, for which his Global crews did much of the on-scene labor. In addition to Global employees, some of whom were "local people (from coastal communities), we had some Washington Conservation Corps workers; then Chempro, which is now Foss, had a crew there," said Davis. (The Chempro crew worked mostly the more southerly beaches in the Grays Harbor area.) "We're about the only two companies now (in the oil salvage business). We have the mega company and the small company, us. We're independent, like Sause Brothers.

"The real key to doing oil on the coast is local knowledge, and we had a problem with ignorance on our part. We...learned a lot from tribal folk on the coast. The women from the tribes out there...are absolutely the best workers we've ever had. Unemployment is tough, and it's an opportunity for them to...make good money. For some reason they just seem to turn and churn a whole lot better," said Davis.

"There's one reason," I suggested. "It's their native land, the traditional we've-always-been-here situation. It's theirs to protect and take care of, so perhaps they approached the work with a different attitude."

"They worked a whole lot harder than the guys, seemed like," was Davis' response. His reference to "guys" meant the regular members of his crew from Seattle.

Kittle commented that Davis is a hard driver of his work crews, won't abide a slacker but is protective of those who perform up to expectations. A tough taskmaster, but one who also cares about his workers.

Perhaps in illustration, Davis had told how Global used the chartered helicopters to commute.

"You see a lot of neat things out there, you're hiking the beaches and you see wildlife, seals, sea otters and all that stuff. It's an interesting place. I've been spoiled by the helicopters, don't think I've walked a lot of beach. We had some hairy times out there when we couldn't get a chopper started, and thought we were going to have to dismantle it or carry it up the beach. We've left them on the beach overnight, even stacked them up on pallets to keep them from washing away. Helicopters are pretty fragile, and we had big problems with that. You get...guys are not experienced working around them, slam doors and stuff. It's not good for them. The door on a Hughes 500 is about $6,000. Or he'll leave his seatbelt hanging out, and of course the pilot can't hear it, he's got headphones on, and the seatbelt beats up the honeycomb outside.

"It's hard to coordinate (your helicopters) out there, because you've got 100-hour checks, your downtime, and it puts you a helicopter short. But they're a great tool. I do know a lot of people don't appreciate them out there, but it just makes our life so much easier. The thrill is gone. In the early seventies, when I was just an oil spill crewman, I used to think, 'One of these days I get to ride in a helicopter.' And now I get up in the morning and think 'Oh, I gotta listen to that thing all day.' Tim (Beaver, Global partner) and I probably spend more hours in the helicopters, almost as much as the pilots. The only time we're not in them is when we're running crews. The rest of the time we're surveying beaches, doing logistical things, so you're always in the 'helo.'

"Once we...don't have to work weekends," Davis continued, "that's kind of a good time for us, because the guys can go out, work five days a week, then come home and be with their families over the weekend. If you're going to have them drive home from Forks (on the west side of the Olympic Peninsula), especially in the summertime with all the campers and tourists, they're...six hours getting home...(at) 11:30 Friday night. And you've paid all this overtime, plus the gas mileage, plus the ferry charge, plus there's the danger...of them being real tired and possibly in an accident. And then Monday when you go back out, you have to catch the first ferry at 5:50, that means they have to be up by 4, and they don't get out there until maybe 11, so you don't get on the beach until 12 noon.... You lose half a day's work. So when you add up all the expenses, it's cheaper to fly. So we bring the 'helos' home to Boeing Field, and they can get their 100-hour checks there. Probably only 6 to 9 or 10 of our crew live in Seattle, when we're using mostly local people, with three 'helos' coming back to Boeing we pay the flying time anyway. So our guys get home early and enjoy the weekend, and then Monday morning we can leave at 6 o'clock and we're out on the beach at 8:30.

"But we got to doing that so much...that I...wouldn't even go to Boeing Field, just have (the pilot) let me off on the beach here (in north Seattle) and the wife would meet me with the car. My mom and dad live in Sequim (near Port Angeles), so they're used to us flying over, and we'd just stop and hover, I'd shove my dirty clothes out the door and they'd fall in the back yard. Then we'd stop on the way back out. My mom's so sweet, I'd always have clean clothes Monday morning if I'd stop, land in the back yard and pick them up."

"There were also the days when the weather's bad and the customer's pushing to get the work done, and the Coast Guard's pushing and all that; but the real person, the 'god,' so to speak, is the helicopter pilot,...and if he says he's not flying today, that's it," Davis concluded.

Global Diving & Salvage also worked on the *Tenyo Maru* spill the summer of 1991, which oiled the northern coast from Cape Flattery southward to Ozette and beyond. "It's kind of ironic," observed Davis. "I don't think Sause Brothers got credit for it, but when the *Tenyo Maru* sank in 550 feet of water, and oil was just oozing out and spreading over the ocean, Mick Lights from Fred Devine (Salvage Company of Portland) was able to get a hose into her with an underwater robot. But the only pump that would move that oil...up to a barge was this little peristaltic pump. Nobody owned any except for Sause Brothers. They saved a lot of oil from getting on the Washington coast. The pumps cost $6,000 each, and Sause Brothers gave those...to us to use on the *Tenyo Maru*.

"Another thing that Sause Brothers didn't get credit for," Davis pointed out. "Absolutely thousands of cubic yards of plastic and (other) man-made debris...was cleaned...off those beaches and paid for by the cleanup. And it didn't make any difference whether it had oil on it or not. If it was something that didn't belong there, pick it up. They should have got some kind of credit for it,...but they got kicked around a lot.

"When Dick Lauer (Sause representative) would leave us in charge,... I'd call him at home every night, report what we had done that day, what we planned to do the next day. If I had a problem...he was there the next day. Company plane would drop him off at Quillayute or Ocean Shores, and I'd have a helicopter waiting for him, and he was on the beach. ...There was a lot of times they may have rolled their eyes, but never once did they try to get us to short cut or whatever, to get...it done as soon as possible.

"Lauer and Rick Kimberley (another Sause supervisor) ran that *Nestucca* cleanup out of a phone booth, and they did a better job than 70 people did on the *Tenyo Maru*. One day I was sitting there,...agonizing over how *Tenyo Maru* was going, and...talking to Dick on the phone, and one of the people from the park said 'who're you talking to?' and I said 'I'm talking to Dick Lauer,' and he reached over and grabbed the phone and said 'Dick, get up here and get this thing straightened out and done for us.'

"He's just an absolute no-b.s. guy, extremely hard to work for, real demanding. He's a full bird colonel in the Air Force (Reserve), but he knows his business real well."

Oberlander of State Ecology, whose account of knee-deep oil on the Quinault Indian Reservation has been mentioned, also made a point relative to other beach cleanup crews that contributed along with Global Diving & Salvage. The Chempro (now Foss) workers were assigned to continue work along the beaches south of Olympic National Park when the Global crew moved north. And Washington Conservation Corps crews worked cleaning both southerly and northern beaches, Oberlander recalled. "Those WCC crews were able to accomplish more than the contractor crews," he said.

Disagreement with Davis' disparaging view of the *Tenyo Maru* cleanup management was expressed by Oberlander, who pointed out that it was conducted under the "incident command system" which involves all interested parties working jointly under shared leadership.

Oberlander spent considerable time in the air during the spill, and accompanied Dr. Galt of NOAA on at least one flight in a twin-engine State Fisheries airplane in search of the main oil slick off the coast. A storm had come through in late December, preventing aerial surveillance, and when clear skies returned, "We flew off the coast many miles, back and forth, and we found a little bit of a slick, and we thought that was the end of it." As Galt explained, the slick couldn't be seen because the oil had broken into globules and spread over a wide expanse. "Yes," agreed Oberlander, "this was a big spill, but it's a tremendous ocean out there, and it required a twin-engine plane to fly that far with any degree of safety. We may have missed it, or because of the loss of the sheen, we saw it, but it was hiding under the water."

Speaking in a more general fashion, Oberlander observed, "In any response, safety is number one, so you're not going to risk people (unnecessarily) in helicopters or in dangerous situations. I got trapped and had to run from the Raft River up to the Queets when we were walking that stretch. It's a long way, and the tide was coming in. We were being pushed to the bank. I was with a young Coast Guard petty officer, and it was a bit of a challenge. We jogged a good part of that beach. I had a radio, but the only way it would have done any good is if somebody flew by.

"Another thing on a more personal note. I think I went through three colds out there (during the *Nestucca* cleanup). I lost 10 pounds, even lost my voice for awhile. I was chairing the 1700 meetings with no voice. My daughter...still reminds me of 'the Christmas you weren't here'."

5

Winter in the Park

Personal, front-line experience with *Nestucca* oil along the beaches comes from an interview with Diane Harvester, an Ecology Department worker. She was assigned as a beach supervisor, and wrote her journal during that time.

"When the first calls came in, I wasn't on duty," Harvester said in an interview at her office. "I started at 4:30 a.m. January 4...assigned to...LaPush from then until 21 February. Actually I was housed at Forks. I would meet the Eagle Air helicopter at the Forks airport each morning for the shuttle to the beaches. Sometimes I would check in at LaPush (cleanup headquarters), but most days I would go straight to the beach, and I'd be the first on and the last off.

"The beaches that I walked...(were) mostly from Cape Alava down to the Norwegian Memorial, plus a few places further north. Usually they would land me on those beaches...mostly just little bitty beaches. I felt...fortunate because most of them are very isolated. I remember one time, after the cleanup was done, the crews were gone, the hubbub had died down, I had been dropped by a helicopter and was...carrying just a small fanny pack, tooling down the beach, and met this backpacker loaded down with his heavy pack, been hiking for days. He looked up at me with the oddest expression. I didn't have the heart to tell him I'd not walked all that way, but was there courtesy of a helicopter. The look of dismay on his face is what I remember; he's out in this remote place to get away from people, then to have me come along...."

Webster: "Of course that's one of the values of Olympic National Park, its relative remoteness and lack of human impact."

Harvester: "Poor Howard (Yanich, park ranger), he was so stressed out during that time about his park, and how horribly it was impacted. He was just a wonderful person to work with.

"We discovered we were flying these beaches with the helicopters real low trying to determine where we had to send the crews. But you couldn't tell (from the air) if the beach was oiled. Once we found that out, my job kind of evolved. It was 'You go down there and tell us if it's dirty or not.' Actually (with practice), I started getting pretty good. I could estimate for Thom Davis (co-owner of Global, the cleanup contractor) roughly how much beach was there,....(the size) crew we might need and what the nature of it was. We had crews that were specialists in removing wood debris, chainsaw 'gurus.' Then we had people that were better at raking and picking through the little debris. It...depended on which helicopter we were assigned."

Webster: "Speaking of these different crews, how many workers were actually on the beach? Global Diving & Salvage had two crews in the park of 12-13 people each, and you mentioned the Washington Conservation Corps."

Harvester: "You know, it was hard to keep track of that, and I started a little log in the back of my notebook...late in the process. It listed what beach, how many crew and what we removed. It varied wildly. One crew down south was 15 people. Then we had a smaller crew way up north, six, roughly, and a medium crew, this was the WCC crew, 10-12 folks. (There were) 32 total that particular day. So that's how it went. I've made an interesting little note here. We talked in terms of slings, because that's how things came off the beach, in cargo nets slung below helicopters. A sling is one (loaded) cargo net, approximately 600 to 800 pounds (varying by the weight of fuel in the helicopter.) So I talked about how many slings of wood, how many bags of debris and sending out the pompoms."

Tribal women on the cleanup crews were "more meticulous," Harvester suggested, adding that they were mostly down at Toleak Point, further south. "The boys I worked with, the conservation crew, were all young men in their twenties; their supervisor might have been 23. It was quite an inspiration to watch them work. And I...watched a couple of boys who were kind of shy and socially uncommunicative actually blossom. These people had a job, they had a mission, and it was exciting to get hauled out there in a helicopter.

"It was quite a thrill to be doing one of the 'recon' flights, and you're down low, close to the deck, and the little stall buzzer, indicating ground proximity, going all the time. I was thinking, 'gee, I hope this thing keeps flying'."

Webster: "Were you still finding live birds?"

Harvester: "Very few. The problem I started to see was the seagulls and what not would come in and start picking at the carcasses, and I'd see secondary oiling on the face and neck and beak. They'd be preening, trying to get that stuff off. It may have killed them."

Webster: "The oil from eating the carrion, and what they preened, might have been fatal because it gets their liver."

Harvester: "I noticed a couple of deer one day down on the beach, and it was disturbing because they were foraging in the seaweed, and I saw they were quite oiled on their legs. The helicopters coming and going scared them away." (Deer commonly eat seaweed in winter, when snow covers much of their usual food supply in the forest.)

"I saw seals almost every day off the beach, but I never found any dead ones. Found one dead sea otter. He was pretty badly decomposed by the time I found him on a very remote stretch. I was out for the day on one of my solo 'recons,' and I didn't have a bag, so I just marked it with flags (plastic streamers) and left it...there, and...ended up going back and getting it a couple of days later because the U.S. Fish & Wildlife folks wanted to see it.

"The birds, bags of birds. After awhile, the guys really didn't want to deal with them, so they'd say 'Hey, come on over, we got some more birds.' I bagged...and tagged them, kept a tally in my notes of how many bags..., how many birds and to some degree what species they were. But mostly it was impossible, just a pile of old feathers and bones.

"One of the things that was really amazing was the chopper pilot from Eagle Air who worked mainly with the crew I was with, his name was Clint. I've no idea what his last name is. Clint is one heck of a pilot. He would drop his hook down just right there for the boys to run up and hook their sling. He would take it back (to the dumpster a mile or two from the beach on a logging road), and one day I found a live bird so I flew in with him to take the bird. I was stranded...for a couple of rounds at the dumpster. He would fly in,...just one fluid motion. He'd let that hook go, and the bag would sail right in and plop!, right in the dumpster. That guy was an ace. It was so darn noisy in that chopper I never had a chance to talk with him. Probably he was an ex-military pilot. He lived in Forks, and I think did helicopter logging. Eagle Air has a hangar there and a couple of choppers. It must be a branch from their main station at Boeing Field (in Seattle).

"Usually I was only half a day ahead of the crew. We really had a system going. I could radio, say 'I got a hot spot here,' and leave a flag marking it so they could fly over and take a look.

"There was a point of rock right above Norwegian Memorial that we started calling Taffy Point, that's where the pompoms were first used. They worked pretty well. I remember at one of the helicopter dumps they'd put a whole box of rope, and then the pompoms were loose in boxes. There were four or five of us on the beach wondering how to clean up this mess, and we started a little assembly line, one guy feeding the rope out and handing us pompoms to tie on...with a simple loop knot.

"Once we figured out this was going to work, Thom (Davis) was particularly proud of the discovery, and it was really neat to see how well those things soak up oil. Those lines were so heavy the next day (when) you'd come back, it would take everything you could do to pull them in. Sometimes we had to cut them in pieces again and bag them.

"It was so cold, you'd be trying to tie those things on, your gloves would be either wet or full of oil or both, (your hands) kind of like clubs, you know, and trying to do this. Life on the beach was pretty darn cold, wet and miserable some days. One day we...could see this squall coming in, and it was pretty bad. It was late in the afternoon, we had to time it pretty carefully because of tides so they (the helicopters) could get in. The helicopter would start taking people off late in the day, and I was usually one of the last. It was getting higher and higher tide and darker. I was really nervous. But he came back and got me as the tide was coming in. He landed very carefully, just touched the toes of the runners to the beach. And I was in surf, about calf deep, climbing on, and the back ends of the runners weren't even touching.

"Another time one of the...pilots, a guy out of Seattle, balanced on a huge boulder and said 'Step off very carefully.' That's hard to do when you have on oily rubber boots. One night I remember trying to climb on in full-dress yellows just covered with oil, gloves dripping oil. I reached up and grabbed the hand loop overhead. It was a real big step for me, and I stepped up, and my foot slipped and my hand came through that loop, and it jerked. I had a mighty sore elbow for a few days.

"I remember being awfully tired of Forks restaurant food, but they packed a really decent sack lunch for us. We'd go in and grab a good, hot breakfast, and with the sack lunch we could last the day. The last few weeks I was up there I moved to the far north end of Forks. It was kind of like (an) apartment...with a small kitchen, and it was great. I'd come in, be so cold, tired and beat. Didn't feel like getting cleaned up, going back down to the restaurant. I was usually one of the few women in those groups, didn't feel like dealing with the guys, you know."

Webster: "Did you get much hassle?"

Harvester: "No, I didn't. It was really excellent the whole time. I think everybody respected the job I was to do. I have to say everyone was very supportive and concerned about my welfare. It was real good. I tried to be real consistent in the calls that I made, so they knew what to expect. And I got better as time went on. A couple of weekends my husband came up to see what I was doing, (but) he didn't see much of me during that time.

"We really did come up with some (new) ideas. I remember we were up near Cape Alava at Sand Point, and the oily debris was...kelp, those big long tendrils (with the bulb at the end). That presented a problem, because it was horribly oiled, we had to get it off, and I'll tell you, it's a real chore trying to stuff that into garbage bags. So do we whack it up, what are we going to do with that stuff? One tendril would fill several (plastic) bags. We solved the problem with bulk visqueen (plastic sheeting). We'd lay out an empty cargo net, spread a piece of plastic tarp (over it) and pile on the kelp.

"Another thing that was really amazing to see (was) the polypropelene rope (washed up on the beach). It attracted oil so it was...obvious. I can't tell you how many hundreds of yards of rope I tagged for cleanup. And even if it wasn't very oiled, I would throw it onto the pile, because it didn't belong on the beach. It's amazing how much is out there. It did soak up some oil, but it was disturbing to see so much of it.

"Somewhere in my notes there's something about how much driftwood we took off. And it was a hard judgment call. I agonized over every single piece of wood. Where do you draw the line for protecting the beach? I got to...where we made a judgment call; if it looked like it was weathering quickly,....I'd leave it, because...the net good of leaving that log for erosion protection was greater than the little bit of oil that we might eliminate from the environment. Some of the logs we would scrape if...(they were) significant to the stability of the beach. If it was just spots (of oil), you could take a putty knife and peel 'em right off. Also we found...that as they rolled and eroded in the surf, they...literally sandblasted clean. I would go back in a few days, and an area that had looked really horrible had actually cleaned itself up a lot. The places that were just thick and covered, a significant threat to wildlife, we took it out. We would...chainsaw it into pieces that would fit in those cargo nets and haul it out.

"Other times we were right down, literally, on hands and knees raking through little bitty pieces of driftwood, and I...would decide, we're leaving this, take this; and the kids were on hands and knees raking, shoveling, picking through this stuff, taking out what didn't belong there or was oiled. That was part of my learning, too, trying to figure out what was significant, 'how clean is clean.' I hope they were good decisions. And that's pretty much how we left it. We left some...., but we got 99%. We did. Considering what we started out with, and how we finished up, I think we did okay.

"I completely wore out a pair of Red Wing Irish Setters (boots). Laminated insole separated from the outer sole. Those boots were just history at the end. I put a few miles on them. Hiked a lot during that time. One time we were hiking out to Sand Point on the boardwalk (from Lake Ozette to the beach), and it was frozen, iced up. Someone loaned me a pair of...crampons. My feet were blistered by the end of that day, because (the crampons) were heavy and I was kind of dragging them. I didn't slip, but it was a six-mile hike that day, three out and three back with those crampons, like hiking with five pounds on each foot. And they didn't exactly fit. That's always been a...problem for women in these...jobs where gear tends to be pretty big. My full dress yellows (waterproof jacket and pants) didn't fit well, either, but you get out the scissors and just whack 'em off.

"Down by Toleak we hiked in once, and...there wasn't a trail (but) we needed to get to this beach and there weren't any choppers available. So we bushwhacked, and...I think the fellow (leading us) was a timber cruiser. I'm pretty short legged, and this man was flying over logs and crashing through devil's club, and I'm wondering if I'm going to make it to the beach. Which way *is* the beach? It was so wonderful to finally break out of this dense underbrush and there it was, we were on a headland. Luckily, choppers were able to pick us up at the end of the day. I kept wondering how soon we should leave to get back before dark, but (on the radio) they said 'We'll have room, we'll pick you up.'

"Radio communication worked pretty well. There were some days when I had to keep it (the radio) inside my coat, in my armpit to keep the battery warm, because it would stop working if it got too cold. I didn't carry a spare battery pack, but usually it would last the whole day, and I used it sparingly. The only time I got on the radio...was when I had something significant to report. Usually I just used rolls of surveyor tape...(to)...flag things. I (would) sketch little maps, and at the end of the day we...(had) debriefing, and I would talk about what I saw, where it was, and (we would) plan the strategy for the next day.

"The...continuity with me being there every day all those weeks was (that) I started to be able to evaluate relative to other incidents how serious this was. My first few days I was going 'Omygod, this is just horrible.' But relatively speaking, it was nothing compared to what I saw later. So...we needed that consistency, and I think it's...critical that...if we ever have--God forbid--another of these things, the crew will be on site for periods of time so you get the consistency. That's what they liked about me being there every day. I was...maybe a pain in the rear end some days, but I'd been there every day, and that was the biggest benefit.

"That was quite an experience. I'm happy that I could participate, and I'd like to think I helped in making a difference. It was time invested that was worthwhile. It was a horrible event, but it made us a lot better prepared. Our spill team made dramatic improvements in how we respond."

The journal Diane kept during her winter in Olympic National Park follows, courtesy of Washington State Dept. of Ecology archives (5-1).

NESTUCCA BARGE/SAUS BROS.

1/4/88 WED [Bunker C Spill] Rain
PER TELECON/Lew Kittle "be in Forks @ 0800"
- 0430 left ELMA Arrive Forks 0730
- 0800 Eagle Air - Pilot Clint Bell 500
    P.O. Terace King MSO Portland USCG
- 0825 Ozette Ranger Station
    Ranger Tim Devine
    WCC Crew - Will, Crew Leader
    - Helo crew (6) and gear to beach
1000 King & Harvester hike to beach
    Arrive @ Sand Point 1100
    • Met Tim Beaver, Global Diving
      "Lots of birds" 50 to 75 ALL DEAD
    ∘ Oil trapped in heavy winrows of
      Kelp - cannot separate so plan to
      chop to bag size pieces and remove
      from So, side of Sand Point
    Walked No. Side of Sand Point
    Many patties - 1 yd² size and
    smaller. Surf grass may have
    covered oil yesterday as none
    was seen on the inspection
    ∘ Checked Rocky tidal areas
      - Some small blotches 25¢ & smaller
      - Some protected pools have light
        sheen

NESTUCCA

1/4/88

NOTES FOR DEBRIEF @ 1700
- Dumpster – should fill current one
  @ Ozette by noon tomorrow
- More Crew to keep up with Helicopter
- Birds – collect and bag separate ?!
- We need to walk Beaches – oil
  is <u>NOT</u> readily seen from air
- Use visqueen sheet in helo sling
  to make one large bag – save time
  and energy by not chopping kelp down
  to bag size pieces

1540 10 Bags birds @ Ozete Ranger
      Station. Approx 60 birds

1715 Quillayute Station / Mtg.
      - Tom Davis, Global : finish 3rd beach
        Tomorrow, move to 2nd
        ~35 dead Birds
        8 to 10,000 lbs removed
      - Tim Beaver, Global : SandPoint
        - 90% kelp on beach oiled
          Need 1 or 2 days more
        - Approx 8,000 to 10,000 lbs removed

1/4/88
- Howard ≏ Park Service
  Beach Walk Teams
    1 @ Shi Shi Beach
    1 @ Ozette to Sand Point
        DOW Sea Otter Survey
    1 @ Norwegian Mem, South to
        Cedar Creek ~2 miles
    1 @ 2nd & 3rd Beaches

Hi TIDE 10:45

- 15,000 more plastic bags on their
    way from Seattle
    (special run made for spill)
    Used 7,000 bags @ Ocean Shores

- Oil reached Canada – Long Beach
    (similar to Ocean Shores) < 5% affected
    ~3 mile stretch   paddy size

- Moclips – saw oiled Eagle
    USFW will try to find him

- Spill estimate between 168 to 230,500 gal
    4,000 to 5,850 barrels at 42 gal ea.

- Clean up estimate - "weeks"
- Parks Dept would like to see more crew. Concentrate on specific Beaches and clean to completion (Be able to Quantify cleanup)

- Discussed Moving command Post to Port Angeles
   Rick Herbert CO from MSO Seattle

- Now a Bird Center at Neah Bay
   Vet & supplies
   50 Birds at center
   2500 Birds at Ocean Shores

Radio - use F2 if possible
   F1 on repeater can use if F2 won't reach.

NESTUCCA

Thrs 1/5/88
   0745 Forks Airport Eagle Air
      • Flew coast - La Push North to Sand Point
   → Noted sheen on beach off Southern
         Tip of Lake Ozette
   • 0800 Met Ranger Kevin MacCarthy
   • 0900 On Beach at Sand Point
         Patties on North side of Sand Point
         100 yds North up to log jam
         - 8 to 10 bird carcasses
   • Noted slightly oiled gull
         Tail feathers affected
         Tar on beak and feet
         Harlequin ducks seemed ok
      [Took ① oiled scoter back to Ozette
      [Will call Neah Bay Center 645-2334
   • Observed gull eating on carcass
         Beak and feet are oiled
   • Two Deer on beach - eating/licking
         seaweed - some oiled.
         Noted oil on their lower legs
   1245 called Jim Oberlander, Ocean Shores
      • North grenville Clean
      • #50,000 requested by Ocean Shores
         for Convention Center
         Must be OUT by 1/26.

Ozette Ranger Station Has Pay!! Phone!

1/5/88 cont.

- Called Neah Bay - will try to meet Ann at Ozette to pick up oiled scoter
- Walked Beach after lunch
  North of Sand Point to Next Rocky
  outcrop. Oil is sparse <25¢ size
  Thin line of oiled seaweed at high tide
  mark.

1600 No. Sand Point Clean (unless more
  washes in)
  - So. Sand Point to Creek - 90%
  will finish tomorrow.

1700 Mtg.
  • Captain Felton MSO Seattle assuming OSC
  • Pacific Strike Team will be here
  - 2 Drums @ 2nd Beach to remove
  Tomorrow: Move 3rd Beach Team to 2nd Beach
            Finish Sand Point - Inspection PM
  - Canada: Barkley Sound ? Long Beach
            Total of 7 days work
  - Hoh to Ruby Beach "Clean"
  - Live Birds (North) Neah Bay
    Dee Burns 645-2334

NESTUCCA

FRIDAY
1/6/89 Sand Point
   Beach @ 0800
      • Some oily surf grass on North Side
        but not much additional oil over night
High Tide @ 1030
   ☆ 19 Birds bagged

• "Polished" South to creek all day
• Inspection at 1530
      — Kevin MacCarthney, Parks
      — P.O. King   USCG MSO Portland
      — Tim Beaver, Global
      — Diane Harvester, DOE
   Sand Point is "CLEAN"
      Kevin will monitor and report to DOE
      any changes in status.
   ☆ Need to work beach South of Creek
1630 Began Lifting Crew

   SAND POINT CLEANUP
      ~14T Debris removed
         80 Dead Birds
         3 Days

SAT 1-7-89

- 0730 Airport Cold 22°/sunny
- Norwegian Memorial Beach with Tim Beaver. Could Not see oil from chopper. Quick landing - found clumps of oiled seaweed (mostly surf grass) Clumps all frozen and covered with frost - Hard to see oil until you turn over clumps.
- Stayed on Beach to continue assessment Tim on to Ozette (North) to organize clean up crew.
- Walked South towards headland

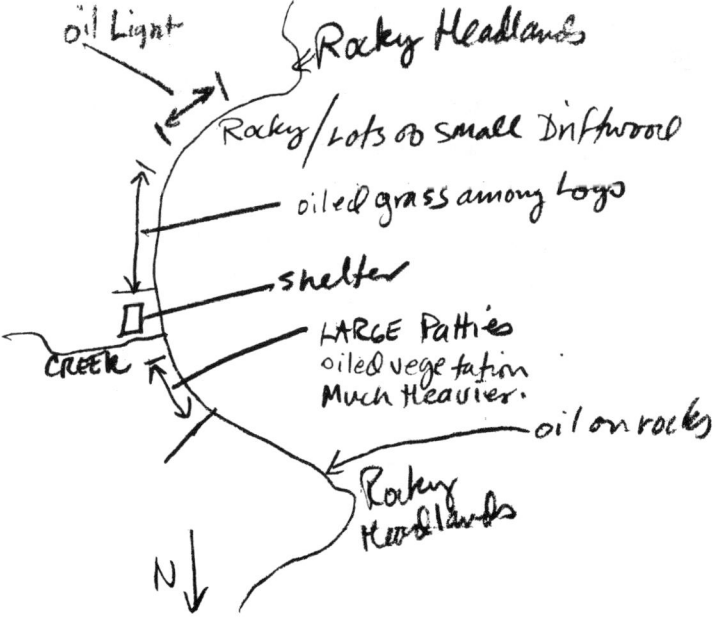

oil Light
Rocky Headlands
Rocky / Lots of small Driftwood
oiled grass among logs
shelter
Creek
LARGE Patties
oiled vegetation Much Heavier.
oil on rocks
Rocky Headlands
N

NESTUCCA
1/17/89 Cont. (Norwegian Memorial Beach)
- oiled grass & small (<25¢) patties
  Random - sparcer as I went south
- North towards creek & shelter
  oiled debris all along tide line
  up to hi tide mark.
  Heavier as I proceed North
  - Stopped at large root wad
    ~ ¼ mile north of creek due to
    high tide
- Per Tim: Looks like 4 Dumpster skow
  (~32T) 3 days work
- Saw <10 Oiled Birds in first walk
  will follow crew and Bag Birds
  - Blue heron - no obvious oil
  - grey gull seems distressed. Some
    patches of oil on head & neck - excessive
    frantic preening.
- Noted sheen in many tide pools @ creek

- 23 Birds collected from headlands
  (south) North to shelter at creek
  Bagged and tagged.

- Finish @ Sand Point

Sunday 1/8/89

0745 Forks Airport   FOG & SNOW
• Clint can fly without us,
  If beach is clear, he will fly
  on to Ozette and get crew onto
  beach @ Memorial
• Tim & I overland to log landing
  on bluffs above beach.
0930 On Beach. Fog & Light rain.
  Crew continues to work North.
  Tim, CG, & I will Fly North to
  rocky oiled beach for recon.
  Hopefully move crew there
  after Hi Tide @ 1207 to clean
  up some of the heaviest oil.
• Checked beach south that was
  cleaned yesterday. Very little
  new oil.

② Birds

1030 • Fly over oiled Rocks @ Norwegian
  Mem. Oily line on rocks clearly
  seen from the air.

1/8/89 Sunday Cont

- Continued flight north to wedding rocks (between Sand Point & Cape alava) Checked oily logs - appear to be 200-400 Large logs heavily oiled. Can't tell how thick from the air

- Crew worked north - reached creek @ Shelter by Noon
  Hi TIDE! 9.9ft @ 1207 forced us off

- Hiked North to rocky area to wait for tide to allow passage around headlands. Command Center folks flew out - saw our position & hi tide and called off rock clean up

- Hiked back to creek @ Shelter Started working area North of creek Heavy oil. Large patties 2 to 3 ft² 3 to 4" thick

- Will leave tools here and continue this stretch in the A.M.

1630 Chopper coming to take us to command post

CREW @ Memorial! 15 WCC
  Will go home tonight
  To office Mon. re: Car & Supplies

Heavy Squalls this PM   1/10/89 TUE

0800 Norwegian Memorial Beach
     Pt. Cloudy cold Strong Wind out
     of North West

• Crew continued work on stretch north
  of Shelter/creek yesterday. All large
  patties picked up — some fresh oiled
  grass washed in today so area will
  have to be reworked

• Walk south of creek to rocky Point
  Marked 3 Hot spots to go back
  and reclean. No Birds

• Checked Tide Pools — Tide low/incoming
  Noted oily surf grass (pictures)
  Sheen on many pools especially where
  rock meets sandy beach (pictures)

• Will continue cleaning north to
  headlands as tide permits. Plan to
  rework oiled washed in materials
  tomorrow

1210 Strong Wind & Hi Tide forces us off
     rocks @ North End of beach.
  → Saw new evidence of oiling on
    rocks as high as 10-12 ft above
    ground level.

1/10/89 Tue Cont.
- Walked back to shelter picked up birds
  total ⑦
✱ NOTE - this tide has brought oily grass
  all along freshly "cleaned" beach
- High Tide in logs forced us off beach

1400 Began reworking beach cautiously
  Some waves still hit logs on upper
  beach.
  - oily grass continues to wash in
    getting mixed into hopeless tangle
    with small pieces of driftwood.
    Especially bad in areas where logs
    hold pockets of debris
✱ NOTE - Hot spots I marked to the
  South this AM have been
  swept relatively clean!

1500 Called Chopper to see if we could
  land at Yellow banks for recon
    "TIDE TOO HIGH"
  will try tomorrow AM

  Wind building - HEAVY Seas / Drenching
  COLD ———— YUK   squalls

1-10-89

• Wildlife today @ Norwegian
  Memorial
  2 Seals
  1 oiled cormorant
  6 Oyster catchers (Looked ok)
  4 Harlequin ~~Ducks~~ OK
  5 ~~Dead~~ Pacific giant Chitons
   (No evidence of oil)
  2 Bald Eagles (I've ~~seen~~
   them ~~twice~~ – they never
   approach the beach or dead
   ~~birds~~)

Neah Bay Bird Center:
  98 Live
  78 Washed
  554 Dead

• Crew @ Memorial: 6 Wa Cons Corp
      (WCC)

1-11-89 WED   Light rain/overcast
              Slight Breeze
        ⟶ BALMY compared to yesterday

• 0745 Eagle Air
• 0800 Nor. Memorial Beach (Kaya stola)
   Crew of 8 today (some off for R²R)
   Will continue clean up North & South of
   creek @ shelter. SLOW process - widly
   dispersed oily surf grass in small drift
   wood among large log jam. New
   material washes in daily — Not a
   great quantity but looks UGLY!
   LOW TIDE @ 0900 Hi 1430
• Tim Beaver, global Diving / Howard Yanish, Parks
   and I flew to Yellow Banks - Very little
   to clean. Checked beach just south
   Found long stretch (> 1 mile?) with
   patties & oiled grass - Enough to recover.
• Stopped at "Taffey Point" (just North of
   Norwegian Mem. Rocks) Met Tom
   Davis working with another crew
   Checked his technique using oil/snare
   Pom Poms.
   ⟶ Oil on Big rocks is VERY Thick —
      apparently refloats at high tide

1-17-89

Strings of PomPoms attract oil.
Each String collected nearly a
barrel of oil yesterday (per Davis)
He will continue to place pompom
lines as long as collection works.
1000 Back to Beach at Memorial
   → Bald Eagles flew over
Walked from shelter South to headlands
   • Highly visible sheen/mousse washing
      in at low tide line, collects in
      tide pools.
   • NOTE — Water that seeps into
      Skid marks made by the chopper
      is "sheeny" Oil trapped in Sand??
   → Watched 4 oyster catchers
      One looked oiled(?) re: "scruffy"
      feathers on Neck & Chest. Didn't
      act sick —
   → Cormorants - looked ok
   → Many gulls - saw no oiled
   • Beach from trail head at South Headlands
      North to First SMALL creek has fresh
      oil - 50¢ size patties oily tide mark
      2 Crew back to rake up.

1-11-89 Cont

- Worked Beach North of Creek
  all PM
      6 sling loads out
      Mostly oiled grass

★ • I scraped logs, used large clam
  shell - worked well on thick oil.
  Put sand on spots and rubbed in
  with back of shell. A stain
  remains on log but oil is not tacky.
  → I think crew equipped with
    putty knives could recover a lot
    from logs and rocks. —
    Idea is ignored or "po-poed" as
    too "tedious" (as if crawling on
    hands & knees picking up blades
    of oiled grass isn't!)

② Birds

⑦ Birds from a volunteer group
  at Cedar Creek
  They left a bag of ~20
  behind that need to be
  picked up.

HEAVY RAIN/<u>WIND</u>! 1-12-89 Thrs
0630 Left Forks to Port Angeles
0800 Met/Laurie Levander DOE
       Jim Beam, Fisheries
       Walter Cook, Fisheries
Re: Beach walks along straits

Laurie/Jim
 • Freshwater Bay — No oil
 • Crescent Bay — Light oil
 • Agate Beach — No oil

Walter/Diane
 • Pillar Point (East) No oil
 • Clallam Bay/Slip Point — No oil
 • Neah Bay — ? "STAIN" on sand
   ~200 X 4 ft strip, Took sample
   Doesn't look like "our" oil
   No Petroleum odor
 • Mukkaw Bay
   South Waatch River — VERY slight
   South of Sooes River — Light oil
   • 3 mile stretch = 12 small 3 to 4"
          patties ? some oiled grass
1630 Report to Bill Young RDA/DOE
   Total of 7 Birds all Day
   → 5 were gulls with <u>Light</u> oil

Fri 1/13/89   Norwegian Memorial Rocks
STORM LAST NIGHT — Heavy wind
Pt. Cloudy 31°   Low Tide   TOTAL 10yd³ Removed
0800 Rocks located just North of Kayastola
Beach.

N ↑   I marked apparent "end" of oil
with float in tree. ~ ½ N of stack

shore line / Cliff
oil all along here mixed in
small cobbles

Large Rocks
to med size Cobbles

(HOT SPOT)   Pom Pom Line

Sea STACK

Clint can land about here
during low tides!

(crew concentrated efforts here today)

Pom Poms →   oiled logs, heavy packet of
thick oil

Large Rocks — packets of oil can
be seen deep in crevices.

Pom Poms →   More oiled logs and pockets of
oil among big rocks

Large Rocks — passage @ Low Tide

old engine →
seen @ Low
Tides   To Kayastola Beach
              ↓ Norwegian Memorial Shelter

⑤ Birds collected   1/13/89

- Small coves either side of sea stack are Heavily oiled. This is the thickest oil I've seen!
- Low Tide Beach only — no passage around headlands at Hi. No place for chopper to land at Hi Tide.
  Beach is backed by rock cliff. Large boulders down to med. sized cobbles. Oil is inbedded in rocks and also heavy globs among drift wood and Logs. Heavily oiled logs here —
- Crew will begin with large, easily recoverable patches of oil first.
- Sling loading small to med size heavily oiled logs
- Bagging small sticks of drift wood
- Raking through small rocks (1 to 3"dia) removing clumps of oil
- All pom poms placed by T. Davis collected oil except for few above hi tide line
- Removing saturated poms — replacing line of poms plus adding two more lines south of sea stack.
1100 - 1300 Tied Pom Pom lines
1400 We've got to get off this beach! Last group off got their feet wet while loading into chopper.

TUE 1/17/89 Heavy winds/Rain last night
★ • Crew could not work yesterday
0800 Eagle Air 37° pt Cloudy /some lite showers
  • Tides won't allow access to Norwegian
    Memorial Rocks until >10:00.
  • Crew driving to dump site at landing
    to load stock pile into dumpsters
  • PomPoms reportedly working well
    all wkend.
1030 Flew over beach — tide still too high to land
1100 Landed — Checked Pom lines
      All completely saturated (~500 poms out)
  • Will load all poms to landing and replace
    with new
  • Two chain saws — will cut up heavily
    oiled logs. Discussed keeping removal
    at minimum. Basically first row of
    logs in heavy oil pockets. Finding LOTS
    of readily cleanable oil pockets under
    these logs!
  • 7 Slings @ ~600#/ea
  • 16 Log pieces @ 5 to 8 ft long
  • 4 Dead Birds
  • 480 Poms tied out
      Note: Concensus — oiled Poms = ~20#/ea
  • To Quilayute Station 1630 Windy ROUGH RIDE
    HEAVY Rain — everything is soaked.

Note from 15th
18 slings
removed &
~150 oiled
Pompoms

WED 1/18/89

0700 Breakfast /Lew, Tim Beaver /Clint (Eagle Air)
- No access to beach @ Memorial until 11 or Noon
- Clint on another chopper job until then
- Tim to Toleak Point per "Clean Sound"
  request: "Beef up crew - speed up cleaning"
  Tim: I will explore overland route
  from logging landing to check feasibility
  of large crew approaching beach on foot.
- Clint will work with WCC crew; continue
  removing oiled logs as time permits
- Meet at Quilayute Station 10:00
- Local "Pete" to guide us on logging roads
  Main line south off La Push Road. Turned
  left on 3110. Took small spur road
  left at Top of ridge.
1130 On Trail. Some flags put out
  by Parks but they were SPARSE!
  VERY ROUGH going - "trail" non existant
  Lost flags - followed tracks down
  slope. Lost tracks. Realized we
  were lost when we came out at
  Junction of Goodman & Falls Creek
  3 mi too far South. Crossed Falls
  Creek found overland trail - hit
  TOLEAK Beach at 1345!

WED 1/18/89 Cont.   NESTUCCA

- Crew of 30 "Locals" plus Global crew of 6 working Northern end of beach near Point
- Sand has covered LARGE patties. Crew looks for tell tail sign - seaweed sticking out of sand.
  Have dug as deep as 10" ave 4 to 5 inches
  NOTE: Heavy sheen/small oil globs leach into water left in excavations
- Some oiled Logs at Jackson Creek
- Oiled cobbles North of Creek at high tide mark
- ☆ These oiled logs & rocks are more lightly oiled than at Norwegian Memorial the oil is not tacky in many cases and has "dried" to a stain

Questions to me — How far to carry patty excav. to Continue with Big Crew here? Remove Logs? Remove oiled cobbles?

Mtg with Lew DOE / John Weyhert - Clean Sound "Continue digging" "Don't sweep oil under carpet"

- ★ Check ITT Rayonier: re burn logs on landing?
- ☆ Experiment with burner to "Melt" oil off rocks
  Toleak clean up report
  ~ 200 Bags of debris
  ~ 500 pom poms @ ~ 20# oil/ea

Thrs 1/19/89 Hike Sand Point/Cape Alava

- Bill Young went along
- Began hiking at 0715
0815 Sand Point
    South 2 creeks (3/4 to 1 mile)
      • Light to Sparse oil
      Small area (20 to 50 ft) south 1/4 mile
of 1st creek south of Sand Point contains
1/2 doz Large patties (6" diameter)
→ ∘ Note these are partially buried
    ✶ More under SAND ??
OVERALL: Insignificant oil visible today
South of Sand Point

- Beach at Sand Point (previously Cleaned)
Sign of re-oiling. 50¢ & smaller size
new (black & shiny) oil. Sparse over whole
point. Kelp debris is quite clean.
→ Low Priority job to "Polish". Needs
short crew (4 to 6) 1 day.
    • A few logs with Light splotches
    of oil. No removal needed.
1030 Break @ Sand Point re: High tide

⑨ Birds

Thrs 1/19/89 cont.

1400 Cape Alava

- 3/4 mile South of Wedding Rocks
  - 100 to 200 oiled logs / some need to be removed
  - oiled sea weed & small wood
  - Patties up to 6" diameter
★ Dead Otter (sea otter) ~24" long
  Badly decomposed. Oily fur

- 1/2 mile North of Wedding Rocks
  (Flagged tree)
  - oiled seaweed & patties - getting buried
  - a few logs & lots of small oiled driftwood
  - Dead Seal
  - Dead Unknown (small seal?)

- Cape Alava "Clean"

- Back at Ranger Station Ozette @ 1530
- Forks @ 1700
  - On to La Push for 1700 mtg.

Rain/Fog   Fri 1/20/89

0800 Quilayute Station
    Tom Harrison — USCG Pacific Strike Team
0930 Beach #3 @ LaPush
    • Hiked North from trail head
      to the rock cliffs
    • Inspected high tide marks
      and log jam all along this stretch
 * In general, the clean-up here has
    been very <u>good</u>.
    We found insignificant traces of oil:
      * Occational small patch of oil
        on logs. Many spots have
        "dried" and/or been ground in
        sand & surf until they are now
        non-tacky stains.
      * < Couple dozen small (<25¢ size)
        patties. Very dispersed along
        large stretch of beach.
        Most likely washed in on tides
        after clean up (re: fresh, shiny
        black appearance)
      * Few (10 to 15) pieces of small
        (<2ft long) drift wood that
        are oiled completely. Still "tacky" oil.

Fri 1/20/89 (cont.)

→ * Found one heavily oiled piece
of rope/net/Floats tangled in
drift wood that should be removed.
  • Require: 2 to 3 men, 4 to 5 bags,
    Cutting tools etc.
  • Location ~ 1/2 way between trail head
    and rock cliffs to north at mouth
    of very small stream. Right on
    leading edge of drift wood windrows.
* One oiled bird (put him in bags of
  birds at trail head)
→ Note: 3 bags of birds to be removed
from under tree at trail/beach junction.
1130 High Tide forced us off beach - returned
to Station.

★ • Clean up @ Beach 3 essentially complete
  * Remove oiled rope/net
  * While walking in for net removal,
    bag up random bits of oil & small
    drift wood.
  * Remove bags of birds
→ 2 to 3 men @ low tide, 3 hours, 1 sling load max.
TOLEAK Rept: 26 Crewmen
50 Loads of wood & 15 slings of debris

1/24/89 Tuesday

Norwegian Memorial Rocks
- Met with: Tim ~~Beaver~~, Global
  Tim ~~Mooney~~, USCG strike team
- Crew continues to remove oiled logs
  and small wood - concentrating
  on area just south of sea stack
  working North.
- Heaviest oil remains on beach
  at sea stack
- Walked North to the next point
  oil tapers off ~ 1/4 mile North
  of Stack. Marked off northern
  extent of oil with orange float & 2gal
  plastic carboy tied in small fir
  tree overhanging beach
- 900 pom poms have been placed -
  2 to 3 parallel lines along heaviest
  oil.
- 1100 high tide forced us off beach

- Plan for tomorrow - remove oiled
  tips of large trees that lay perpendicular
  to beach (just North of Stack)
  - remove logs & debris as needed

1/24/89 Tue Cont. (Norse Mem. Rocks)
- Goal is to clear rocky beach of
  logs etc to aid in enhancing surf
  action. Will continue to set out pompoms
- Large accumulations of oil will be raked
  up and bagged.

N↑

clean

oil tapers off about
here ~1/4 N of stack
Tagged with float in tree

Light oil- couple of "pockets"

Large oiled logs

pocket of oil
and wood debris

SEA STACK

Heavy oil -
pom pom lines
here

Large Rocks
some oil in pockets

oil & oiled wood
(Heavy log & wood removal here)

HEADLANDS

TO
Kavasbola

1/24/89

- •1430 Flew to Wedding Rocks
  - Crew of 7 working South of Rocks
  - Found otter carcass from Thus. Hike:
    Turned over to Ed Bowlby, Dept Wildlife
  - Discussed oiled logs to be removed
    from beach south of current work site
    * Marked oiled stretch with floats
      in trees : showed to Rich & Tim/global

N↑   WEDDING Rocks

crew worked here today:
oiled logs, debris, patties
should finish Tomorrow

Rock with Hole in it

Found Otter

oil Logs (2 to 3 days?)

Marked with Blue plastic Bottles Hanging in tree

HEADLANDS

CLEAN

SAND POINT

WED, NESTUCCA
1/25/89 Memorial Rocks
FOG/Mist & Light Rain
0800 Walked South to Shelter @ creek
on Kayastola Beach
- Needs "polish" work
- Found some patties partially buried
in sand. (just North of creek
- Some "New" oily debris on tideline
- Marked 5 hot spots on Memorial
Rocks. (Orange Tape with white float)
NOTE: Pom Poms are not collecting
oil re: High tides are too low to
agitate the heavy oil line
- Crew continued to remove logs that
are "holding" pockets of oil
- Strung 13 Boxes (2390 poms)
More tomorrow if supplies arrive
☆ - Dug into clean looking small (2") cobbles
Found oil down ~ 4 to 6". Possibly deeper
if I had a shovel ———
1130 Tide Forced us off beach

Norwegian Memorial Rocks   1/25/89

N↑

SEA STACK
← gridwork of Pom Pons
Heavy oil in Large cobbles

LARGE TREE
Big Boulder

Pom Pom lines

BIG LOG

① Flagged
oiled Driftwork
Most bag size

② Flagged
Oil in Large Rocks

LARGE Jagged Rocks

③ Flagged
oil in rocks
some Logs

Rock "Fin" Cliff

④ HOT SPOT / Flagged
oiled Logs
and driftwood

⑤ Hot Spot / Flagged
pockets among
LARGE ROCKS

HEADLANDS

Engine buried in Sand

4 Bags on log
orange Float

Kayak fold
Bench

Buried Patties
here

3 bags on logs
Yellow Float

CREEK

Shelter

1/25/89   Wedding Rocks
  1200 Noon
      - Crew working just South of Rocks
      - Oiled logs - (few)
      - oiled driftwood/debris - bagged
      - Beach looks good where clean up
        crew has worked
          Should be done by tonight as far
          South as "window" Rock
      - Walked North of Wedding Rocks
        • Found oil ~ 1/4 mile North
        • Approx 1/2 mile stretch affected
        → oiled kelp - large masses
        → Patties - large "banks" of oil embedded
          in coarse sand and small gravel
        • Hard to determine quant. until crew
          begins digging . . . .
★ • Marked affected area with survey Tape
   Begins at Large Stump North of wedding
   Rocks / TAPE and Orange w/ Black stripe float
   continue north from stump — several
   Hot spots flagged (TAPE in trees)
  NOTE: More oil here than I remembered
        from trip thru this stretch last week!
  1500 • Took Tim BEAVER i Rich — (Global)
      showed them stretch ——

MON 1/30/89

- 0400 Left Elma
- 0650 Arrive Eagle Air, Forks Airport
  No one at hanger — went into town
  couldn't find anyone at restaurant.

0720 Roger from Eagle Air - "Chopper
  down for repairs" no flight today

0800 Quillayute station -
  Marion @ trailer - new home for
  wkend. Thinks he will be back
  today. Bob (NWRO) filling in
  as OSC.

  - Discussed wkend activities - No work
    Sun. due to weather
    Wind starting to pick up now - Forcast
    for High winds, rain turning to snow.

0930 NORWEGIAN MEMORIAL Rocks
  Strong wind - ROUGH RIDE !!

  - WCC crew of 8. Nothing taken off
    beach re: no chopper support.
    - 5-6 Sling loads of bagged debris
    - Some logs cut & ready to lift
    Working area ① see map 1/25/89

1230 Pulled off beach. Heavy rain,
  High wind, No visibility. Storm is Here!
oiled areas look much the same - still in
  rocks. Poms not working re: No High tides yet.

Tue 1/31/89
  0730 Eagle Air, Forks
       Snow, Fog - Can't Fly
       Drove to La Push Command Center
• Burn pile at Toleak -
       - Began burn Mon. A.M.
       - Good progress - Expect to be finished
         by Wed 2/1/89
       - Road access - not good due to snow
         and general poor condition of roads
• Discussed sample plan for all dumpsters
    re: try to determine % oil in debris to
    calculate gallons recovered.
       - Problem in obtaining accurate samples
         due to wide variety of debris containing
         widely variable amounts of oil.
       • Looked at feasibility of composite samples
         Would involve sampling large # of bags!
       • Began some best guess calculations
         with Tom Davis, Global and John Wiechert,
         Clean Sound.

98.2 Ton oiled Debris bagged
~ 100T oiled logs to burn

252,000 Max gallons Spilled
~ 20% evaporates
0.2# oil / 1# debris = 2% oil @ ~ 7.5#/gal

<u>Gal. Recovered</u>
Ocean Shores   1,000 gal
Sand Island   376 gal
La Push   450 gal
Ozette   250 gal
* Taffy Point (PomPoms) 8,000 gal
Whale Crk Dump   500 gal
                 _____
                 10,576 gal

* ~19,000 PomPoms recovered
  Ave oiled wt 5 to 10#
  (as high as 25# ea)

2/7/89 Tue   Clear, Sunny, Frosty
• SAND POINT INSPECTION • with J. Weathers USCG
~ ½M South @ Small Creek / old boat hull
• Marked small area Blue Flags - drift, some
   Spattered logs. (Hard to see extent due to frost)
   Area between Flags ~100ft or so
• ~150 yds North - found oiled seaweed
   under logs. Spattered drift. Left bag
   of rocks & drift   One Blue flag high on Beach
• 50 ft Between two flags
   oiled rocks baseball size & smaller
   some small drift wood
• ||| Small patches proceeding North
      Blue flag. Spattered rocks, drift &
      logs (burn off oil ?)
   • Between two flags - cache of bags oiled debris
      area needs more work
• 2 Oiled Logs (Big) well off spatters
• 2 lg oiled logs + couple of smaller pieces
• oiled logs - splatters, some dollar size
   patties
• oiled wood 5 or 6" chunks ~6ft long
—CREEK
   • North of creek ~½ dozen logs with splatters
Flag • Two splattered logs
Flag • Black Tarp covering oil, 5 gal Bucket + bags of oil ?

2/7/89

- 4 Bags oiled Debris Just south of rocky Point
2.6 Bags debris just North of Point
- ~6 Bags 150yd North of Point
- Oil splatters on Sand Point rocks are not weathering. Try to melt off?
- Sand depositing on point area re: buried seaweed mounds

S ↑

Fiberglass boat Hull

11 areas flagged with Blue Tape
~1/4 mile stretch
Spattered rocks and logs

Creek

X ← spattered logs
X ← 5 gal bucket/ Black Tarp covering patties

CLEAN!

X Bags

SAND POINT

X Bags
X Bags

N ↓

2/7/89 Cont. NESTUCCA
Norwegian Memorial
- Kubota tractor "helo'd" down.
  Did not work — <u>no</u> traction!
- Global put a 3 wheeler and a make shift
  "scraper" down. Used it to open up one
  of the oil-berms above high tide mark.
  Limited success — managed to dig down
  to bare dirt. Oil has permeated the
  entire gravel layer in spots. Approx.
  3 ft deep ave 8 to 10 ft wide — 3 or 4
  spots like this.
- Global & CleanSound looking into bringing
  a Cat to the area.
* Objective is to scrape oiled bank
  down into active tidal area, cover with
  network of pompoms and collect oil
  as tide flushes it out.

Clear, Sunny, Cold   Wed 2/8/89

- NORTH WEDDING Rocks
  Jim Weathers, USCG Strike Team
  Debbie Davis, Global
- Inspected beach approx 3/4 mile north of Rocks
  Flagged all hot spots
  ★ Lots of oil has been buried by small
    gravel (1" to 2"). Some gravel will be removed
    to recover oil.
- SOUTH WEDDING Rocks
  · Flagged a few hot spots south to
    "Hole in Rock" point. Area needs some
    polish work
  · Logs south of "Hole in Rock" point
    are looking quite clean! Lots of sand
    and gravel has washed in among the logs.
    Very few visible oiled logs — WCC crew
    from Memorial will work this stretch
    and hopefully finish today
1430 North wedding rocks —
    Most flagged areas now clean
    NOTE: 41 sling loads off this stretch so far
    — Crew very close to completing.
1700+ 58 Slings off Nm Rocks
    WCC crew found Hot spot in logs — will finish This.
  Rocks for final Insp. Fri

2/9/89 Thrs.   Yellow Banks
- Flight to Headlands just North of Banks
  Walked North to headlands just south
  of sandy beach before Sand Point
  - Tim Beaver, Tom Davis, Debbie Davis - Global
    Lt. Kirk Smith, USCG   Harvester DOE
  - oiled debris, small rocks, few patties,
    some splattered lg. rocks
    Flagged area - Pink
- Checked crew progress @ Sand Point
  - Picking oiled rocks north side of point
  - WCC cutting up oiled logs south
  - Global short crew finished off loading
    slings of material at N. Wedding Rock
- Flew to Cliffs just South of Yellow banks.
  - Walked area with D. Davis, Lt. Smith.
  - Oiled wood, debris etc. starts at rocky
    point just south of Yellow Banks Bay
  - At Large, heavily oiled tree (perpendicular
    to beach) a thick bed of gravel/oil
    appears. Traced heavy band of oil
    for 300 to 400 yds South. Dug band
    in several spots - oil 6 to 8" deep all
    along. Gravel is small (~1")
    LOTS of oil. Called Tom & Tim to view.
  - Finished hiking south to rocky headlands
    beyond Two Bit (Hammock) Beach. Flown there

Fri 2/10/89

0800 LaPush
- flew to Oly. Bob Penhale from CRO here.
- Briefed Bob on current activity of crews. Made
  copies of maps - high lited current Hot Spots.
- Bob to do survey flight with Tom Davis &
  Sause Bros. Reps - Rick Kimberly & Dick Lawer

• WCC & Global crews to finish beach south of Sand
Point and start on Pockets We flagged yesterday
North of Yellow Bank. WCC crew may start
Log removal Yellow Banks <u>south</u> this P.M.

• Prepare Map for Jim O.
     Mark Hot Spots, Clean Beaches Etc.

• Crews found more oil than expected
  on southern end of Sand Point Beach.
  Will finish early tomorrow

2/11/89 SAT.   Clear : cool
- Inspected <u>Kayostda</u> w/f T. Harvester
  - North end at 1<u>st</u> Rocky Headlands:
    * Pocket of oiled Rocks & driftwood
    ~10 bags of debris (Yellow Flag & Orange
    float) on large log
    * 3 areas south (Yellow flags)
      oiled logs & smaller wood
      < 10 logs to cleanup total
    * Just North of Creek (Yellow flag)
      4 to 6 bags of debris

    * <u>South of creek</u>: orange float in tree
      • Rake, Shovel & Bags of debris (3 or 4)
    * Orange Flag in tree:
      2 bags debris - oiled rocks - (3 or 4
      bags? to pick)

2/12/89 Sun   Cool, Overcast, Drizzle

• Inspected 2nd Beach @ La Push
  — South end of ~~Beach~~
      cobbles along high tide line
      are oiled, Tend to be heavy
      but not sticky stain.
  ☆ Using Lauer's criteria of
      < Volleyball size > 50% oiled
   there's about 30 yd³ to
   remove (SWAG !!!) spread along
   ½ mile stretch of beach.
  — Not much wood — a few logs
   one flagged — Most could be
   scraped or burned off.

  ☆ NOTE small cove far south end
   of beach LOW TIDE access
   ~30 feet wide around a
   small creek. Oiled cobbles &
   some small spattered drift wood

2/13/89   Whale Creek Dump Site

Hwy 101 South – Cross Queets River
~1.6mi so. of Queets look for gravel
road off to right (west) Marked with
blue paint on tree 4700. Road
goes west ~½ mile then turns 90°
South. Cross two small bridges
(~2½ miles) look for dump site
just beyond 2nd Bridge on right
side of road in small turn out
area.
  – Est. 4 to5 pick up loads of
  bagged debris.
→ Dump truck would be quicker
    IF bridges could take weight

• Checked beach at mouth of creek
  Clean

• Discussed Grenville and Raft
  River with Vicki Watkins 276-4424
  in Tahola.
• She has seen some oiled debris
  at grenville but is not sure of Raft.
  Does not know how accessible
  Raft is by logging roads.

Log Piles - Burnables   2/12/89
• Bob Penhale inspected for approx size

Ozette Station (3 piles)
  ① 30′ × 30′ × 6′ = 5400 ft^3 = 200 yd^3

  ② 30′ × 30′ × 6′ = 5400 ft^3 = 200 yd^3

  ③ 32′ × 45′ × 4′ = 5760 ft^3 = $\underline{213\,yd^3}$
                                 613 yd^3

CEDAR CREEK 5600 RD

210′ × 36′ × 4′ = 30240 ft^3 = 1120 yd^3

2/14/89 TUE   Morning Clouds, Clear PM

BEACH CLEANUP PROGRESS REPORT   1630 hrs

LOCATION	STATUS
• Shi Shi	Cleaned 2/8. Ready for Interim Inspection
• Wedding Rocks North	Cleaned 2/9. "
• Wedding Rocks South	Cleaned 2/9. "
• Sand Point	Cleaned 2/12. "
• Yellow Banks North (Pocket Beaches)	Cleaned 2/12. "
• Yellow Bank	Cleaned 2/12. "
• Yellow Banks South	Cleaned 2/12 ‡ 2/13. Oil debris ‡ logs removed. PomPom lines set.
• Two Bit Point	Cleaned 2/13. Ready for interim inspection
• Norwegian Memorial	Oiled debris ‡ logs removed. Oiled cobbles ‡ gravel remain above high tide line. PomPom lines are in place and will be monitored as needed.
• Kayostla	Cleaned 2/14. Ready for Interim Insp.
• 2nd Beach	Cleanup in progress. Estimate completion by 2/15
• 3rd Beach	Cleaned ‡ Inspected. Ready for sign off.
• Whale Creek	Clean. Removal of dumpsite required.

LOCATION	STATUS
• Raft River	Very light "spatters" found on log and rocks. Weathered, not tacky. Not feasible to persue cleanup.
• Point Grenville	Lightly oiled small debris mixed in small drift wood. Require short crew with rakes to "polish". Need to walk beach South of Point.
• Sand Island	Small tarballs on west end. Require short crew for approx. one day.

2/15/89   Cool; Clear, Calm Sea Very Low Tide

Interim Inspections
    Bill Lehr   NOAA
    Howard Yanish ONP
    Kirk Smith   USCG
    Tim Beaver   Global   623-0621
    Dick Laver   Sause (503) 269-5841
    Diane Harvester DOE   753-2353

• LOCATION   COMMENTS

1. Shi Shi   Need inspection 2/17 ?
2. No. Wedding Rocks   Flagged one spot for polish
3. So. Wedding Rocks   Flagged 5 spots: oiled rock, scrape Logs
4. Sand Point   ⎰ Good - Scrape two big rocks at point
                              ⎱ Monitor Pom Poms for So. End
5. Pochets No. Yellow Bank   Monitor Pom Poms
6. Yellow Banks   Needs interim inspection
7. So. Yellow Banks   oiled debris removed - Pom Poms in place
8. Two Bit Point   Monitor Pom Poms
9. Taffy Point   Monitor Pom Poms
10. Kayostla   Ready for Final Sign off
11. 2nd Beach   Pom Poms southern cove
12. 3rd Beach   Ready for final sign off

Dumpsites   2/15/89

*OZETTE STATION
- Log Pile (~610yd³): Move to ITT gravel
  pit for burning. Waiting for bids.
- Remove 2 dumpsters of oiled debris
- Oiled rock (~25yd³)
  Tim Holth, Del Hur Const. Port Angeles
  will take rock for batch plant operation
  <u>if</u> DOE approves of interim storage

* CEDAR CREEK 5600 ROAD
- Log Pile (~1120yd³) Begin burn 2/16/89
- Dumpsters for PomPoms (will remain as
  long as needed)
- Oiled Rock (~2yd³) DEl HurConst?

* LA PUSH
- 4 Dumpsters oiled debris
NOTE: As of 2/10 182,780 lbs debris to landfill

* WHALE CREEK
- 6 to 8 pick up loads oiled debris to be
  hauled to La Push. (~½ done today)

OTHER ITEMS
- Long Term/Pom Poms: Howard will watch tide
  and storm sequence. Will call Global.
- Final Inspection/Sign off next week
- Dismantle La Push center 2/17 PM?

2/21 Sand Island, Ocean Shores

• 1400 Met Tom Davis, Global Diving at
Ocean Shores Airport and P.O. Miller, MSO Seattle.
- Sand Island clean up will take
remainder of week. A lot of oil
has been uncovered as sands
have shifted. No New — all old & weathered.
Flew out to island. Crew of 6
working. Choc Chip up to 50¢ size
patties in 8ft wide band along
southern shore of island — oil
tapers off as you go around tip
of island to east. Oil in "wash overs"
- Crew will be increased by 5 starting
tomorrow.
- Plan to complete clean up by Fri.

• Toured dump sites
- Airport - 6 slingloads today 46T still active
- ButterClam - ~180 yd³ (40T)
✱ Batch plant rejected material due to
other oiled debris mixed in.
Will take this material to dump at
Aberdeen.

Other activity:
  ① Dumpster removed from La Push
-Global discussed 30yd³ oiled rock
  with Del Hur Const.
  They plan to used "oiled" rock
  in ratio of 8 Tons to 20,000T total.
  This crushed mixture will most likely
  be used by Summer due to road
  project near Sequim.
→ Need written approval from DOE

Tomorrow:
-Remove another dumpster from La Push
-Crew of 3 global unload "Heavy" dumpster
-Take rock from 2nd Beach - consolidate
  with rock at Ozette
-Crew to grenville in truck. When
  Chopper finishes shuttle to Sand Island,
  will lift sling to truck. Global
  plans to drive straight to dump at
  Aberdeen.

NESTUCCA

Sling = 1 cargo net to chopper
~600 to 800 pounds each

[Crew & Debris Removal Summaries]

1-21-89 Toleak Point 15 crew - 32 load of wood
Wedding Rocks 6 crew 50 bag debris
Norweg. Mem 11 crew Set out pompoms

1-22-89 Toleak: 16 crew 60 loads of wood
Wedding Rocks: 6 crew 6 bag debris
Nor. Memorial 12 Crew 20 loads wood
15 bag debris

1-23-89 Toleak - Final clean up Insp. on 25th
Norw. Memorial No work re High Tides
Sand Point - Scheduled for final cleanup

BEACH	# Crew	# Loads Wood	# BAG Debris	POM POMS
1-24-89 TOLEAK	12	←—50—→		—
Norse Mem	8	8	—	5 slings out
Sand Point	9	12	—	—
1-25-89 Toleak	11	1	1 sling	
Norse Mem	11	6		5 slings
Wedding	7	?	?	?

Beach	crew	Sling wood	b=bag s=sling debris	~20#oil ea Removed PomPoms
1-26-89 TOLEAK	3	—	2(b)	—
Norw. Mem	9	15	—	(set more out)
Wedding R.	12	40	10(s)	
1-27-89 TOLEAK		~120 PomPoms set out		
Nor. Mem	12	(no count on removal)	6(s)	1 sling
Wedding	12	—	65 bags	—
1-28-89 Norw. Mem	12	set out 20 Boxes of Poms		
		30 Sling wood		
Wedding Rocks	15	(no reports)		
1-29 Norw. Mem.	12	—	50 Sling	set out 30 Box
(North) Wed. Rocks	9	—	600 Bags	—
(South) Wed. Rocks	5	35	—	—
Yellow Bank	4	—	12(s)	—
1-31 Norw.	7	no reports		
Wedding	6	"		
Sand Point	4	"		
Yellow Bank	4	"		

NESTUCCA

CREW ÷ Debris Removal Summaries				
BEACH	Crew	Wood	Debris	Pom Poms
2-1-89 Norw. Mem	7	(down early re SNOW)		10 slings
2-6-89 Wedd. Rock	11	— no report		—
2-7-89 Norw. Mem	?	14 cobble ÷ Rock	41 sling	2 sling
So. Wedding			8 "	
No. Wedding Rock	?	1 sling	25 (cobble) slings	
2-8-89 Nor. Mem	?	10 slings wood ÷ cobble		
So. Wedding	8	10		
No. Wedding	23	58 slings @ 600 to 800 # ea mix of wood, cobble ÷ debris		
2-9 So. Wedding	8	42 slings (finished!)		
No. Wedding	?	22 slings (rock)		
Sand Point	23	75 slings (wood ÷ debris)		
2-10-89 So. Sand	7 WCC 22 Global	97 slings	47 (rock)	
2-11-89 Sand Point	12	—	1 (rock)	—
So. Sand	31	10	20	set out 15 Boxes
Yellow Bank	?	—	10	—

2-12 Yellow Bank (North)   8 sling Rock   set ont 20 Box Poms

"   South   6 slings wood & debris

Nor. Mem   2 Box Poms

Beach	Crew	(sling) WOOD	(slings) Rock/deb	Poms
2-13 So. Yellow	25	13	15	10 Box set
Kayostola	7	137 LOGS	22 slings (wood)	
Two Bit Point	?	3	2	10 Box set

2-14 89 Yellow Bank 3 (globue) set 12 Box Poms

Kayostola   31   14 (sling) 3 (slings)

25 Logs

6

Tales of Oiled Birds

Seabird victims of the spilled oil reached the Grays Harbor ocean beaches 12 hours or less after the hole had been punched in the barge *Nestucca*. The oil that first came ashore with the birds before noon on Dec. 23 had leaked early in the morning while the barge wallowed in an ebb tide with heavy seas outside the Grays Harbor bar, a couple of miles from the beach.

The Daily World in Aberdeen carried the first story on the spill (6-1), an unhappy prelude to Christmas. The banner headline account quoted Lt. Cmdr. Ron Larsen, operations officer at the Astoria Coast Guard station, predicting that the 18-mile slick would come ashore somewhere north of Ocean Shores. Although it was not known at the paper's 11 a.m. deadline, both oil and oiled birds were already on the beach.

The second-day story Saturday (6-2) reported that approximately 50 oil-soaked birds had "washed up or floundered ashore, according to the State Department of Ecology. Rescue teams were being mobilized...in an attempt to save them. Officials...(said) the impact on marine wildlife is likely to worsen." David Wasson, the reporter, quoted a spokesman for the State Division of Emergency Management, Mark Stewart: "The latest reports...indicate an average of 12 dead or distressed birds every quarter of a mile along the beach."

Already, two wildlife rescue groups had set up emergency centers in Hoquiam: Western Wildlife Rescue at Lions' Park and Volunteers for Outdoor Washington using Our Lady of Good Help Catholic Church, the paper reported. "Grays Harbor Sheriff Dennis Morrisette was among the first to discover an oil-soaked bird Friday. Shortly after, in his role as county emergency services director, he declared about a mile and half of beach northward from the North Jetty closed...to assess the damage."

The first photograph was published that day. Larry Waldon of Western Wildlife Rescue was shown holding a white-winged scoter up to an electric heater. It might help you to entertain something of the feeling of that scoter if you think of it sleeping peacefully offshore, bobbing on the waves in its warm layer of insulating down, then being rudely awakened as frigid seawater mixed with oil seeps through to leave it floundering, cold and near helpless. It's comparable to someone dumping a bucket of oil-polluted ice water onto your bed while you sleep.

By Monday, the day after Christmas, more than 1,000 birds had been affected by the oil. Temporary spill-management headquarters had been established at the Ocean Shores City Hall. "It's not nearly as bad as it could've been" said Ron Holcomb, spokesman for the State Ecology Department. There had been fear the oil would be carried into the harbor and pollute sensitive Bowerman Basin National Wildlife Refuge near the airport west of Hoquiam, where tens of thousands of migrating shorebirds stop to rest and feed every spring and autumn. Another danger was that had the *Nestucca* been abandoned, it would have been swept onto the North Jetty or the beach and lost its entire 2.94-million-gallon cargo, not just the contents of one tank. What was not realized until later was that oil would drift north to Vancouver Island in British Columbia, Canada, and even beyond.

The people of Grays Harbor, faced with hundreds of oiled birds, responded immediately to the need for help. The gymnasium of Hoquiam Middle School was "transformed into a makeshift rehabilitation center," Wasson reported.

Santa Claus having disappeared along with toy commercials for another year, people in other communities, seeing on television and reading about the oil spill's effects, began to respond in a manner closer to the real spirit of Christmas. Hundreds, and eventually more than 2,500 volunteers, responded to the need to rescue those distressed victims of humanity's destructive treatment of the natural environment. A photo by *World* staffer Kathy Quigg was widely published. It showed Grays Harbor County Commissioner Bill Vogler about to catch an oil-soaked murre on the windy, rain-swept beach. That is the sort of visual stimulus to get one up from an overstuffed chair and into action.

At the middle school gym, a couple of dozen volunteers were cleaning birds in a mild detergent solution under the supervision of Alice Berkner. Executive director of the International Bird Rescue Research Center in Berkeley, California, the onetime nurse has become one of the world's foremost experts on saving birds from the effects of oil spills. She had been summoned from her home in Vancouver, Washington, by Sause Brothers Ocean Towing Company. In addition to hiring Berkner, Sause Brothers had 25 workers from Global Diving & Salvage of Seattle on the beach collecting clumps of oil and oiled debris for disposal. Another crew, young people from the Washington Conservation Corps, was also on the scene, doing both bird rescue and beach cleanup. The work was to continue for six weeks or more.

While the bird rehabilitation effort was getting organized, the drama was on the ocean beaches. Stephen Clutter of the *Seattle Times* wrote the following account from Ocean Shores (6-4).

"Holiday visitors to the state's most popular stretch of beach found a dismal sight.... Dozens of oil-soaked birds stood dazed in the frothing brown surf. A duck near the edge of the water panicked as one man approached. Trying to fly, the tarred creature flapped its wings. The effort was in vain. But help was on the way.

"While many people finished up last-minute shopping, teams of volunteers spent Christmas Eve trying to rescue the distressed birds. By yesterday afternoon, at least 150 birds had been rescued from the beach just north of Ocean Shores. Between 75 and 100 dead birds were found, according to Jules James of the U.S. Fish & Wildlife Service. James said...volunteers came from as far as Seattle, Portland and Poulsbo.... Most of the birds were...murres, grebes and scoters. Only one seagull was reported coated with oil, and some gulls were eating other species that were dead on the beach. 'Based on past experience we think about 25 percent of them will survive,' he said....

"As ornithology students at The Evergreen State College in Olympia, Eric Larsen and Scott Richardson generally spend their time studying birds. This weekend, however... 'We came here on a winter birding trip and found ourselves...a disaster,' Larsen said...as he and Richardson combed a stretch of beach.... 'Most of the distressed birds probably were caught in the slick offshore, then carried by the waves to the beach. But that's an ominous sign. I expect a lot of them would sink and die at sea, so we're only seeing an isolated but...vivid example of the damage,' Richardson said."

In a letter published in the *Seattle Times*, Richardson pointed out that "one aspect...has been overlooked: effects of oil on the region's shorebirds. Of the 500-plus birds being treated at Hoquiam by volunteers, perhaps a dozen are sandpipers, with a similar proportion of this family among the dead. This...is misleading and causes a biased understanding of the complete disaster.

"On Dec. 25, during a survey of sandpipers and plovers along a 12-mile stretch of beach north of Ocean Shores to Ocean City, 11,500 shorebirds were counted, 3,500 (fully 30 percent) of which had easily detectable oil on their bodies. Like seabirds with oiled feathers, these birds are exposed to the risks of hypothermia and increased predation. They use more energy to preen their plumage, ingesting oil in the process. It is expected that most of these oiled shorebirds will perish prematurely.

"Capturing shorebirds is difficult. They are land birds--not clumsy ashore as murres, scoters and grebes are. They are highly mobile even when soiled, so have dispersed over the entire Grays Harbor region and beyond. Dying birds seek shelter in dune grasses, where search-and-rescue workers will have trouble locating them. They should be remembered, (along) with more than 1,300 seabirds, as victims of an unfortunate incident," Richardson's letter concluded (6-5).

By December 28, some 900 live birds had been collected for feeding and cleaning, while 600 had died, according to the *Daily World's* staff and Associated Press sources (6-6). "Oil-coated waterfowl have been reported as far as 45 miles north and 110 miles south of Ocean Shores, some in areas inaccessible to rescuers," the Wednesday story said.

At the Hoquiam Middle School gymnasium, wildlife rehabilitator Ron Ambuehl commented, "this was a wonderful place to start, but it's not the place to finish." Volunteers kept working at feeding and cleaning birds from 7 a.m. to 11 p.m. in two shifts, the newspaper was told by Kirsten Addy, a Tacoma woman helping coordinate the workforce. "These are people from all walks of life, not just a bunch of college kids on break," she said.

The news reporter visiting the school wrote of "hallways ...stacked...with boxes containing oily birds brought in overnight, and clean boxes that will eventually hold more birds. Newspapers and plastic tarps cover most of the gym floor as some of the hundreds of volunteers rush about. Down the hallway--waxed just before the birds took over--is the boys' locker room where the birds are cleaned.

"Workers...have been able to clean from seven to ten birds per hour in the locker room, but that could increase to 15 an hour with more room and better quarters at the Ocean Shores Convention Center," said Pam Miller, a State Department of Ecology biologist. As she spoke, plumbers and construction workers under a local contractor, Dave DeSoto, hired by Sause Brothers, were swarming through the main halls of the center, cutting plywood panels into 2-by-8-foot pieces for sidewalls and erecting pens in double rows throughout most of the 9,114 square feet of available area. Sliding partitions separated a smaller hall with 1,200 square feet to serve as administrative headquarters, which had been temporarily using the Ocean Shores City Hall.

The operations center moved Tuesday, December 27, and the next day volunteers transported the birds to their new quarters, loading scores of boxes into trucks for the 18-mile trip.

By Friday, December 30, the bird center had received 1,569 live birds. That day wildlife officials released 11 cleaned birds at Damon Point on the south end of the Ocean Shores peninsula, facing Westport across the Grays Harbor entrance. However, because oil continued to appear on the beaches, and after some of those birds first released were soon picked up re-oiled, the much-photographed event turned out to have been mostly for the media. Subsequent bird releases were delayed until January 6, and moved from the ocean beach to the south end of Puget Sound at Luhr Beach near the mouth of the Nisqually River and the National Wildlife Refuge.

Several preliminary problems had to be solved when the bird rehabilitation center was moved from Hoquiam to Ocean Shores. The first was to provide the quantity of water to wash hundreds of birds. Municipal officials installed three 2 1/2-inch fire hoses from nearby hydrants, snaking them across the parking lots and protecting them with heavy planks so vehicles could cross without damaging the hoselines. That provided enough water, in addition to the building's normal supply, for both the bird washing room and for the outdoor pool where cleaned and dried birds were placed to learn whether they would float rather than sink. Even a tiny leak in the outer feathers of a waterbird ruins both flotation and insulation. If water intrudes, the insulating down soon resembles a mattress left out in the rain.

A second startup necessity was to heat the water so birds wouldn't succumb to hypothermia (getting too cold) during the lengthy washing process. The propane heaters installed initially could not do the job, so more heaters were added until there were eight in all, according to Miles Beach, municipal utilities supervisor. Through January the bird center used from 30,000 to 50,000 gallons of water per day, Beach said. That was half as much as the 100,000 gallons received daily by the sewage treatment plant, according to Beach, so it was handling one third more than normal.

All those temporary water heaters at the Convention Center almost caused a disaster, recalled Jim Oberlander, spill response team leader for State Ecology. They were mounted too close to the outside wall, which caught fire from the heat. "Luckily, somebody was out there and got an extinguisher on it," said Oberlander. "After that we moved the heaters off the wall." Another aspect relating to water was whether washing birds left too much oil in suspension for processing at the sewage plant. "Miles Beach needed a laboratory sample," Oberlander said. "It was the day the governor was down, so I cleared it with the State Patrol security people, and we sent the oily water samples back to Olympia in the governor's airplane. We learned that so much water was used, the oil content was below the maximum standard, so we didn't have a problem."

A mystery that showed up after about a week was more difficult. Birds that had been washed, dried and appeared fully rehabilitated proceeded to sink in distressingly high numbers when placed in the test tank outside. Figuring out the problem took awhile. It developed that Ocean Shores water, supplied from several wells about midway on the peninsula, is "hard." Although rainwater is soft in its original state, it soaks into the sand and percolates into the underlying gravel, leaching minute quantities of chemicals from the river-born and glacial deposits. These leachates created the problem by leaving residue on the bird's outer feathers that normally provide the moisture barrier for flotation and insulation. "Microscopic analysis of bird feathers...indicated that grains of mineral from wash water... were breaking their watertight integrity," according to a report by Bryn Beorse in the *Daily World* (6-7). "Then it took until Friday to confirm that even a swim in tanks full of city water outside the center was causing trouble...so the birds couldn't float."

When the problem was identified--one not understood when encountered previously in bird rehabilitation from oil spills--chemical water softeners were installed in the supply for the bird-washing room. That worked, although quite a few birds had been put through repeated washings before then. Washed birds were tested thereafter not in city water, but in the surface water of Duck Creek near the Ocean Shores firehall. They passed.

Alice Berkner, looking back in a 1993 interview, recalled working with Dr. Jessica Porter, a veterinarian helping at the bird center, and Oberlander to solve the hard water problem. "That, I think, was the biggest thing we learned out of the *Nestucca*," said Berkner. "We had run into it before, but we had no idea it was (caused by) the hard water. We go into a spill situation now, and the first thing we do is test the water. Calcium carbonate is what causes the problem. But if the water's too soft, that's also a problem. Now we know exactly, between 30 and 50 milligrams per liter is the best range for birds. That was a...valuable lesson. We knew the water was hard, and...kicking it around,...we said 'let's wash some feathers in soft water and some feathers in hard water and find out. So Jessica, our vet, got someone at the University of Washington, and in their microscopic photos we could see all this stuff."

Since the float testing pond outside the bird center used city water, a different situation was required. "When Jay Holcomb (Berkner's assistant) found the stream (by the nearby city hall), that was a godsend," recalled Berkner. "He said, 'I think we can work with this,' and sure enough, we could. The stream feeds into a canal leading to the harbor. People were terribly cooperative, and it all comes together."

Beorse, *Daily World* news editor, had written a feature that described the washing process prior to solving the hard water problem (6-8).

"'You little sweetheart, you!' exclaimed Jack Rayment of Port Orchard as a badly-fouled murre suddenly struggled for freedom after a bout of dangerous quiet in the soap bath.... Rayment's murre was a hard case. Most birds had been pronounced clean after scrubbings in seven to eight tubs of water containing one percent Dawn detergent. This bird had faced 16 buckets, said Rayment, a service and parts manager for a new car dealership in the Bremerton area.

"His partner in washing the murre, Barbara Cook of Olympia, said it was 'either come here or go to the Nordstrom's sale,' and she plainly felt she'd made the right decision. The penguin-like bird tugged and jerked as his captors maneuvered for more secure holds as they talked.

"A retired couple from Seattle found the pretty grebes more high strung than the strong, feisty...murres. 'You feel like you're dealing with a delicate creature,' said Joanne Terry...while wrestling a murre. Her husband, Bill,...was using a pressure water spray to clean the...feathers of detergent. 'We love animals and nature,' she said. 'It's a wonderful chance to get to know them and learn about them.' The Terrys had come because of the seabirds they see while sailing...on Puget Sound. 'Once you've handled the birds, you never forget which species you're looking at in the wild,' she added.

"At the next tub...worked Masumi Matsuda, a Japanese student of transportation and business at Highline Community College, Seattle. A native of Osaka, she nodded eagerly when asked if she is interested in nature.

"It's a frenetic pace, and Alice Berkner...is the person in charge. Sometimes you wonder if she'd like a break, just like those overstressed birds. But she's always in demand, called away here and there to solve problems, sometimes to the accompaniment of loud avian squawks and gatherings of anxious helpers. The bird center...seemed to be functioning like, uh, a well-oiled machine. 'It was built by a local contractor to our specifications and his imagination,' Berkner said. That's worth a smile,.... But when she notes that a horned puffin, the first one she's worked with, has been released on the north Olympic Peninsula with a clean bill of health, one suddenly senses what gives her joy. Sause Bros....is paying for cleanup work, but it's not money that brings her into this effort. It's plainly love for living things--and a sense of fairness to other life on this planet.

"'We all use oil in some form or another, and human beings make mistakes,' she said. 'I think we have to be prepared to clean up the messes we make'."

Another report on Berkner (6-9), explained that she "doesn't expect to leave...without a few scars. But that's nothing new. The short white lines of scar tissue are a symbol of the past 17 years.... "I don't mind' said the 53-year-old...woman who likely will be putting up with abuse (from the birds) at...Ocean Shores...for at least another two weeks. 'If you look you can see scars from other times' and other places over years she has spent rescuing birds caught in oil spills.

"The self-taught bird rehabilitator's...hands are a reminder...how terrified...feathered creatures are when handled. 'Sometimes when I go...into a store...someone will ask 'Were you working with a wildcat?' It was in 1971, Berkner recalled, that 900,000 gallons of oil spilled in...San Francisco Bay. Some 7,000 birds were affected. Trained as a registered nurse, Berkner was home with her children then and decided to help a friend...cleaning the oil-coated California birds.

"'I had a credit card to that oil company and I felt it was the responsibility of those who use oil to pitch in,' she said of her first experience.... There wasn't a training crew to help,... (and) 'you couldn't buy a book or go to a zoo and just ask how to care for an oil-soaked western grebe at that time. We used mineral oil to clean the feathers,' she said. 'That damaged the feathers and meant nine months in pens for birds to rehabilitate themselves.'

"After that 1971 oil spill...it was apparent to those concerned that (more) oil spills were...the future. With a few others, Berkner helped spearhead creation of the International Bird Rescue Research Center at Berkeley. She and her assistant, Jay Holcomb, travel to where (they're needed). 'I've lost track of the number of spills and the number of birds,' she said, noting (however) that she has dealt with...90 species. 'I'm on call 24 hours a day and I pretty much have been since 1972.' When not responding to spills, Berkner said she and Holcomb are often at training seminars across the country teaching first-response bird cleaning crews.

"She also has a contract with the Alyeska Pipeline Service Company...(of) the Alaskan oil pipeline." That last was, in its January context, prophetic of events two months later in Prince William Sound, Alaska.

Meanwhile at Neah Bay, on the Makah Reservation near Cape Flattery at the entrance to Juan de Fuca Strait, volunteers from the mostly Native American community began finding oiled seabirds on the beaches Saturday, December 31. The *Peninsula Daily News* in Port Angeles carried a Neah Bay report by Kevin Patterson (6-10).

"Oil-soaked survivors lay quietly in the cardboard boxes that served as their hospital beds at the Neah Bay police station, awaiting one of the eight daily feedings they needed to stay alive. Bobby Rose, Neah Bay's animal control officer, stepped carefully among the boxes, talking softly to the frightened grebes, ducks and loons as she checked the yellow stick-on labels on each box to see which bird needed to be fed next. A urinary catheter served as a feeding tube for the Gatorade that Rose forced down the birds' gullets. A gruel of fish food and fluid was being prepared...to feed the birds. Rose was tired but upbeat. Of the nearly 50 birds in her care, only one had died. In grim contrast was the stack of dead birds wrapped in plastic bags outside.

"Volunteers reported seeing flocks of birds, unable to fly but unwilling to be captured. Frightened waterfowl resisted many rescue attempts. Only birds too exhausted or too befuddled to fight or flee were easily taken.

"Before the first oiled bird was found Saturday, Rose had no training how to care for the(m). She called...Dee Byrne of the Northwest Raptor Center in Clallam Bay for assistance. 'She pretty much had to have a crash course from what little I knew,' Byrne said. 'She's been phenomenal because she's been able to do...and absorb so much in such a short time.' Byrne is helping coordinate the bird cleanup. The oil-soaked birds were moved from their temporary home at the...police station to a fish processing plant owned by the Makah tribe, where there is more room and better facilities to clean the birds.

"On the ocean beaches west of Neah Bay, volunteers...Ann Renker, Alias Rose, Bobby's son, and Wilson Arnold slogged through wet sand and a biting wind...to look for more victims of the oil spill. Renker spied a dark shape on the sand of Tsooyes Beach. It was a grebe that looked more dead than alive. Renker knelt to inspect the bird and then gathered it in her arms. She carefully held the bird's long sharp beak between her fingers and set off for her car. Further down the beach, 10-year-old Wilson and 12-year-old Alias found another grebe and brought it back to the car.

"A steady stream of volunteers arrived at the police station, some carrying survivors and others lugging buckets or bags of dead birds...."

New Year's Day papers (6-11) carried reports of a sad event at Ocean Shores, where 350 oil-soaked birds were given lethal injections to put them out of their misery. "They picked birds that were in poor condition and...had a poor chance of survival in the wild," Mark Stewart of the State Department of Emergency Management told reporter Gail Greenwood from Aberdeen.

As of Sunday, January 1, the totals were 2,432 live birds received, 2,175 dead birds received, 350+ birds expired at the center, 277 birds washed and 27 birds released. A carpenter was building additional pens for birds waiting in cardboard boxes.

Bird washing and feeding continued, in an atmosphere described by *Seattle P-I* columnist John Marshall (6-12): "Their hands and arms bear the scars of trying to do what they can to help the birds caught in the oil spill. Gloves are supposed to be worn at all times for protection, but gloves do not always work. The water is too soapy, the oily birds are too slippery and washing them is too tricky.

"So the volunteers have gashes on their hands from the birds' beaks and rashes on their arms from immersion in water. And still they keep at it, amid this maelstrom of soap suds and water spray, eight hours or more of standing on concrete and wrestling with wild birds that are hurt and dangerous.

"'These birds go for the hands or the face or, even more likely, the eyes, so you must know where their beaks are at all times,' Kirsten Addy is telling some new arrivals. 'Do not risk being blinded in this humanitarian effort....' Addy's hour-long orientation...is...their first glimpse of what resembles a major military operation, a kind of M.A.S.H. unit for injured birds.... What is most amazing is that it has gone on for days now, as the terrible wildlife toll...has slowly become known.

"And every morning," Marshall continued, "this command center starts out like a play with an 8 a.m. showtime, but an uncertain cast, or maybe no cast at all. At 6:30 a.m....an Ocean Shores police officer was the only person in the building, having stood watch since 11 p.m. when the center was closed to allow the stressed birds time to rest. By 7 a.m., 140 volunteers had shown up and the center had begun to hum once again."

Marshall observed that "this is not a place for people who live their lives by the percentages, who figure something is worth doing only if it is pretty certain to have the desired effect. This is a place to take the plunge, to work like hell and don't look back and hope it makes some difference somehow. So Louise Smith and Joan Berkoff of Seattle and Bob Demar of Renton spend an entire eight-hour shift on gruesome morgue duty, logging in dead birds and wrapping them in plastic bags and stacking them in a corner. They try to keep focusing on how scientific study of these dead birds will matter later. Doubts do intrude among the volunteers, no matter how committed. All this effort, they know, is being expended to save so few birds.

"Lisa Miazga of Marysville has spent several days sticking a long tube down the throats of birds,...feeding...their mixture of Gatorade and fish pellets. She has seen people of all kinds and ages come together, with this intense spirit of purpose and selflessness. And yet Miazga can't stop...thinking how strange human beings are.... 'It's like,' says Miazga, 'let's run to the rescue and we'll all feel good.'

"Dave Brastow of Seattle has some of his own doubts. Yet he cleaned birds after the oil spill near Port Angeles in 1985 and he is back.... He works Dawn detergent into the birds' feathers with a firm touch, cleans around their eyes with spray from a Water Pic, scrubs their beaks with a toothbrush--a process that takes a half hour for each bird. And the other day, Brastow was half way through washing one bird when he noticed it had died, right there in his hands. 'If it acccomplishes nothing else, being here raises your environmental consciousness,' he says. 'You see what we as a species are doing to our beaches and to our wildlife'."

Doubts aside, all that effort did pay off. And if Marshall took a grimly realistic view of the operation, others could be more upbeat, as was feature writer Karen Reed of *The Herald* in Everett (6-13). Her experience, subjectively recounted, was perhaps more similar to an average volunteer's memories--although one could hope not in such olfactory fashion.

"There I was last Saturday, gripping the steering wheel at 4 a.m. and praying I'd survive the icy roads and make it to Ocean Shores in one piece. I was tempted to turn back a few times, but figured I was already committed. Helping with the bird cleanup was something I was determined to do.

"Why? You can bet I asked myself...on the 3 1/2-hour trip.... I've always considered myself something of an environmentalist. But that wasn't the big motivator. I guess I was more than a little tired of always being the impartial, unfeeling observer, in other words an automaton with a pad and pen. I'm exaggerating some, although I was really beginning to feel that way. I wanted to get in the game for once.

"So when I arrived safely a few minutes before 8 a.m...I was eager to go to work. Other people must have felt that way too, because the worker sign-up area was jammed with people in old clothes, rubber boots and slickers just like mine. I heard later that 580 volunteers showed up, many more than had turned out during the week. It was certainly heartwarming to see all these folk--teachers, secretaries, engineers, fisherman and others--standing around sipping coffee and waiting for their call to duty. But my vision of playing Florence Nightingale to all those oil-soaked seabirds was fading quickly.

"Oh well. As a volunteer coordinator named Bruce told us during orientation, the birds come first. Besides, washing the...birds was only one of many jobs that had to be done. So, when a man picked me and six others out of the crowd for one of the first jobs of the day, I felt like a kid who'd finally made the basketball team. I should have taken heed when he described the task as a 'dirty job,' but I was blinded by my enthusiasm.

"A half hour later, the spirit was still willing, but the body, the stomach specifically, was weak. Lois, a volunteer from Everett, and I found ourselves crawling underneath clean-bird cages retrieving fouled (no pun intended) newspapers and replacing them with new ones. I was comforted, however, by the knowledge that I worked for an industry which produced a very versatile product. While the job may not have been the most glamorous...it was nevertheless necessary. It wasn't an easy task either. For obvious reasons, we were instructed to keep our mouths shut and our rubber gloves pulled up high when reaching under the cages. And we had to move quietly so as not to disturb the already stressed-out creatures.

"If you detect some humor in my writing, please know that I don't wish to make light of the tragedy caused by last month's oil spill. It was important for us to maintain our sense of humor to keep on working. Never in my life have I seen anything so sad as all those once-proud and beautiful birds now soiled and huddled together in wooden pens. I was ashamed to be a human being....

"When the day drew to a close, I was happy that I'd had the chance to contribute to the cleanup effort. As I walked toward my car, I paused near the tanks, where clean, recovered birds were checked to make sure all the oil had been removed.... It was already dark, the ocean air was cold, and I...had a long drive ahead. But still I stayed. The birds swam and...flapped their wings. And my heart soared."

Perhaps Karen Reed was able to put in words the feelings of quite a few volunteers among the thousands who helped out during those weeks when people did what the birds couldn't do for themselves.

The final result of "all that effort" was the release of 1,027 rehabilitated seabirds back into their normal environment. That was believed to be more birds than had been rehabilitated from any previous oil spill. Not only that, but an unusually high proportion of birds brought in alive survived to be released. And all of those birds were released with U.S. Fish & Wildlife Service bands on their legs, so that if one is found in the future, the band can be reported and added to the records.

Bird operations at Ocean Shores concluded January 26, and the few birds remaining were taken to the Hoquiam rehabilitation center. Juli Bergstrom of the *Daily World* reported (6-14) that the bird hospital's odor had dissipated, and quoted Donna Buck, secretary for the Ocean Shores-Grays Harbor County Visitors and Convention Bureau: "You don't smell anything but paint."

Contractor Dave DeSoto had repainted the inside of the cavernous building. The contract covering the agreed-upon $1,019 daily rental by Sause Brothers for use of the center during January had not yet been finalized, but most of the money had been paid. A few events scheduled during January had to be changed, but after February 9, operations at the center were back to normal.

A Tuber is Not a Couch Potato

Nothing appears quite as deserted and empty as a resort area in the off season. That seemed the situation as I wheeled our pickup camper into the community of Ocean Shores on the north spit of Grays Harbor in early January. As usual, it was gray and raining, a brisk wind sweeping off the ocean and across the mostly-leveled sand dunes of the 1960s development.

For a century and more, this area had been used for cattle ranching, originally to provide beef for pioneer lumber mills on Puget Sound. The cattle were driven up the Chehalis River Valley using ancient Indian trails and later the pioneer Hicklin Hill road over the Black Hills, thence to lumbering towns such as Port Ludlow, Olympia, or the growing village of Seattle.

But aging cattle rancher Ralph Minard, after three generations of his family had raised beef cattle among the sand dunes and scrub pines of the peninsula, sold most of the land in the late 1950s to a group of developers, and Ocean Shores was the result: motels, vacation homes, a marina and eventual incorporation as a town. Ocean Shores had been invaded by the tidal wave from the 1964 Alaskan earthquake, but assault by bunker-C fuel oil is of a different sort. Oil on the beach is human activity gone awry, and part of our response comes from knowing that it ought not to happen.

Tuesday morning, January 3, 1989, the news included pleas for volunteers to help care for and clean thousands of oiled seabirds found on the beaches following the *Nestucca* oil spill. With the Christmas/New Year holidays over, students and teachers who had been helping were heading back to class, along with others who could take no more time away from work.

A telephone call to the bird rescue center verified the situation for me. "Still need volunteers?" I inquired.

"All we can get whenever we can have them," replied the voice on the phone.

So I tossed a sleeping bag and a container of homemade granola into our camper, filled the gas tank and headed down the road into a driving rain. Two hours later, as I swung off the main street into the Ocean Shores Convention Center, rain was falling even harder.

This place was anything but deserted, although it had been converted only a few days previous into a bird "hospital" and headquarters for oil spill response operations. The big reader board at the parking lot entrance read, "Oil Spill Command Center. Bring Birds to West Entrance. Thank you."

The parking lot was full of cars, pickups, motorhomes, all manner of vehicles. I found a spot beside one with a bumpersticker new to me: "Avenge yourself. Live long enough to be a problem to your children." Its wry humor caused me to wonder if the owner might be another retiree/volunteer.

At the main entrance doors to the building hung a hand-lettered sign: "Use other door." Mystified, I turned and slogged across the lawn to the west side of the building, where a plastic above-ground swimming pool had been set up. People were standing watching something in the pool, and as I approached there were barricades and ropes, firehoses snaking along the ground, and birds--a small flock of murres--huddled on a platform at the side of the pool. Parents and children were watching them from outside the barrier, while inside the ropes a woman also stood watching the birds. She wore a foul-weather jacket, its hood pulled close to shelter her face from the wind. She also wore waterproof pants and boots.

My camera was hanging on its strap around my neck, so I took a picture of the scene, then asked the woman her name, and why these birds needed to be watched. She was Jody Drew from Seattle, and "I'm watching for sinkers. These birds have been washed, but sometimes one of them can't float. My job is to save it from drowning if it sinks and gets stuck by the current against the outlet screen." That made sense, because a hose running fresh water into the tank kept a steady overflow stream running out. I learned the plastic tank was one loaned by the State Department of Natural Resources from its fire-fighting equipment.

"Only clean-bird workers can come inside here," Jody told me in response to my inquiring how I could get inside the building. "You should be able to get in the main doors at the front." Not quite understanding what "clean bird workers" meant, I trudged back to the main entrance, tried the other of the two double doors and realized that the hand-lettered sign was simply intended to prevent someone from using the door that would catch the wind and send a blast whooping inside.

The lobby was crowded, standing room only, and my glasses quickly steamed up as I worked my way through the people toward the right, where there were a couple of long tables with folks lounging on folding chairs, chatting over cups of coffee. Others, lacking chairs, wandered about, and a few were using a roped-off stairway to the second floor as makeshift seating. Along a sidewall was an urn of hot water for making coffee and tea, and above it the wall was plastered with mostly hand-lettered signs. One provided some totals on birds brought to the center, but was a couple of days behind time.

Other sheets listed various bird species brought in: scoters (white-winged, surf and black), western and red-necked grebes, common murres, loons of three species (common, arctic and red-throated), plus one or more each of canvasback duck, Cassin's auklet, dunlin, pigeon guillemot, gulls (unspecified), horned puffin, black-legged kittiwake and rhinoceros auklet.

Particularly intriguing to me was a sheet inked by different contributors with a succession of wry and awful puns:

> *Don't be loud, be demurre.*
> *If you don't know, wing it.*
> *Try not to make grebeous errors.*
> *Stay happy--don't get down.*
> *It was just murre-der.*
> *Come at 7:30 for the oily boid special.*
> *Murre-aculous.*
> *Murre of the same.*
> *These volunteers look pooped.*
> *And they brought him gold, frankincense and murre.*

That last one, I thought, is well seasoned, considering it's just past Christmas. You get the general feeling, which seemed to express the prevailing spirit of the volunteers and other workers.

Wandering back toward the entrance, I discovered I'd missed the sign-up tables because of my steamy glasses. There was a poster on the wall: "Welcome to the world's largest birdbath, bed & breakfast." Each person signing in was instructed to make out a first-name tag on masking tape to stick on shirt or jacket. Quite a few people, I noticed, had their labels stuck on black plastic garbage bags worn as protective jackets, with holes for head and arms.

If anything could be said to typify the oil spill and bird rehabilitation operations, it would have to be the plastic garbage bag and its use for a multitude of purposes. That, and the combined odors of oil and bird excrement floating on the air should remind any participant forever after.

There was a second instruction for newcomers, which was to wait for the next introductory briefing lecture. About 150 people had checked in already that morning, I was told. Before long, when about a dozen new arrivals had filled the available chairs in the west end of the lobby, Ted Willcox showed up to introduce us to the operation.

He was simply another volunteer, Willcox explained. From Olympia, he was a retired miltary officer, not so much an expert as just someone who'd been around long enough to learn the answers to most of the questions. More recently I learned that Willcox had been a White House communications officer during the Kennedy presidency.

The birds at the rescue center, more than 2,000, were mostly murres, western grebes and scoters, plus some loons, which he explained are more sensitive and difficult to care for, so they were being handled separately. The others were penned by species, all the scoters brought in the same day in one pen, the murres found that day in a second pen and the grebes in a third.

Willcox patiently answered queries from hopeful volunteers, explaining that at the moment there were no jobs lacking workers. Because of the need for relative quiet for the birds, only those people involved in working with them were allowed out in the main area of the center, among the bird pens. We volunteers waiting assignment should stick around, have another cup of coffee, and something would come up soon, he suggested.

So we did, and it wasn't long before a fellow came through the lobby asking for volunteers to work outside. Maybe we'll get to search the beach for oiled birds, I thought; but no. He wanted a crew to police the grounds around the building. Half a dozen of us were willing, and he handed each a black plastic garbage bag and showed us out the back door of the building. On the way we passed the kitchen where volunteers were mixing stuff in huge kettles with warm water. Although somewhat fishy smelling, the odor reminded me of mixing mash with warm water to feed our chickens back when I was a kid.

Outside, the weather had not improved. The wind tried to grab my plastic bag away, and the pieces of styrofoam and candy wrappers I picked up didn't weight it down enough, so I added a handful of gravel from the driveway as our group scattered around the building and parking areas. Cleaning the grounds, hardly helpful to oiled birds, would at least contribute to appearances, and perhaps emphasize the responsible aspect of the bird rescue center's operation. By the time I'd collected as much litter as could be found in my area, the wind was working the chill through my wool shirt, and my ragg wool watchcap was soaked with rain. Time for some warmth.

Inside, a lunch counter operated by Red Cross volunteers was serving sandwiches and a good clam chowder. Somebody said the chowder had been contributed by Skipper's, a seafood chain. Good for them, I thought, and it sure helped warm this chilled volunteer.

So far, I'd had no contact with the actual birds that were the focus of this operation. We could only peer through tiny windows in a couple of guarded doors into the main hall of the center, where we saw dozens of plywood pens occupying almost the entire floor area. The last time I'd been at Ocean Shores Convention Center was to visit a photography exhibit. Now the activity inside was limited to a few garbage-bag-garbed people moving from pen to pen. No birds could be seen behind the low walls.

While eating lunch, I found a friend from Olympia, Donna Van Kirk, also seeking volunteer work. We joined a group of others mashing newspapers and stuffing them into those ubiquitous garbage bags, the paper to be employed as absorbent litter in the bird pens. Just about every volunteer, it seemed, helped on that endless chore at one time or another. Crumpling newspaper makes little demand on your brain, so conversation thrived. Other volunteers had come from all over Western Washington, and a few from Oregon and British Columbia, all attracted by the chance to help save these thousands of innocent seabirds from society's pollution of their winter gathering area off our coast.

Finally, as the 4 p.m. feeding hour approached, the "dirty bird" feeding crew assembling in the lobby came up short one person to serve as a "holder." The crew leader, a chunky young lady wearing the label "Laura" on her garbage-bag jacket, asked if I had experience. "Well, I've caught a lot of chickens, and some baby oystercatchers and seagulls."

"Okay, let's give you a try. This is Mary Gropp. She's a 'tuber' and needs a 'holder'." As we trooped down the hallway to the kitchen, I learned that dark-haired Mary was from Orcas Island in the San Juan group of northern Puget Sound, and was here with her husband, Steve, a commercial fisherman in summer.

At the kitchen each tuber picked up a plastic jug filled with warm slurry, and a 60-cc. syringe with a plastic tube about 18 inches (45 cm.) long. Then the feeding crew moved through double doors into the main auditorium, stopping inside to pick up a little bottle of eyedrops on a loop of twine to hang around the neck ready for use. As time went on at the center, a paper cup was taped to the plastic jacket to hold vaseline ointment.

In the main room, which is more than 9,000 square feet (810 sq.m.) with 19-foot (5.75 m.) ceilings, the odor of bunker-C fuel and bird droppings was more intense. Plywood pens, each eight feet (2.44 m.) square with sides 24 inches (61.4 cm.) high, stood in double rows across the expanse of floor. Each pen had a movable plywood divider, and as the feeding crews started work, other teams of pen cleaners had just finished. The "pooper scoopers" as they were called, had herded the birds in each pen into one side using the divider, then removed dirty newsprint litter and put fresh paper in place, covering it with a clean bedsheet in the emptied half of the pen.

Then the feeding teams went to work, one tuber and holder assigned to each pen of birds. When a team finished feeding the birds in one pen, they moved to a pen holding unfed birds. As a pen of birds was completed, the feeding team moved a "flag rag" from the divider to the pen's outside wall as an indicator. Later on, the cleaning crew would return to do the remaining half of each pen.

Taking it by detail, the feeding process was as follows. The tuber usually stands just outside the pen, sets down the jug of slurry and proceeds to fill the syringe with the prescribed quantity of liquid food. That was usually 30-35 ccs. for murres or grebes, and 40-45 ccs. for scoters, depending on their size (7-1)

Meanwhile, the holder climbs into the pen with the birds, perhaps 35 to 40 murres, for example. They're an unattractive gaggle of black and brownish-gray, some brooding or half asleep, others with beaks in the air and beady dark eyes alert. Those white breasts that make murres look so like penguins, especially when they stand upright, were smeared with brown oily grime. The holder's job is to catch a bird and hold it for tube feeding. Moving slowly and supposedly with deliberation, despite the nervous excitement of a first-time catch, the holder reaches toward the nearest of the huddled birds, hands poised to grab it.

This is simple, you think, until suddenly that closest bird struggles over the top of the crouching murre next ahead of it, and you have to choose a new target. The birds seem to begin moving all at once, and you must be quick and at the same time deliberate to pin down a slippery, oily bird in the smelly, soiled mass of crumpled newspaper. Many a first-time holder has difficulty grabbing a bird, and soon an entire pen of murres can be in frantic movement. You'd like them to stay quiet, but cooperation from the birds is nil. Sometimes you have to back off and wait for the birds to settle down before trying again. It's important not to get the birds overly stressed; but if you have trouble at first, eventually more confidence develops.

When you get a bird, grab it firmly but not too tightly, and pick it up, holding both your hands over its back and wings to prevent flapping. Watch out for the pointed beak on that long neck. A grebe or murre can zap you with its beak as quickly as it catches a fish trying to escape underwater. By all means, don't get the bird too close to your face. (Plastic goggles donated by Weyerhaeuser were worn by a cautious few.)

If you're not wearing gloves, you probably vow to put some on at the first opportunity. I preferred a pair of fitted leather work gloves brought from home, although there was a supply of plastic waterproof gloves of the type used by the bird washers. Leather provided more protection than plastic, which tore when assaulted by beak. There was a first aid station staffed by volunteer nurses and emergency medical technicians, who dispensed not only bandaids but tetanus shots for wounded volunteers. They did a lot of business--unpaid, of course.

But let's get back into the pen with the birds and the holder, who by now should have a murre in hand. When you feel ready, stand up and turn toward your tuber--who's been waiting patiently, we hope. It's often helpful to tilt the bird's head downward to keep it quieter, perhaps because that makes the bird think it's coming in for a landing, someone suggested. But once the tuber grasps the bird's head in one hand, you'll want to hold it up at a convenient working height.

First the tuber will usually check the bird's eyes and put drops in them to help wash out any oil, then will pry the beak open so as to insert the feeding tube. That tube must travel down the bird's gullet and into the stomach, at the entrance to which resistance is sometimes encountered from the cardiac sphincter. But that's less a problem than the risk to the bird's life should the tube accidentally go down the windpipe to the lungs.

If slurry is accidentally pumped into the lungs, the result will be a dead bird within moments. A similar outcome may result if the tube is stopped short of the stomach, which can cause the slurry to back up and overflow into the air passageway and lungs.

"Watch for the neck feathers to ruffle as you slide that tube down," is the endlessly repeated directive of the feeding supervisors. It's the only sure way to avoid going down the wrong route. In the somewhat dim light of the Ocean Shores auditorium, however, it was often difficult to see those neck feathers ripple, especially on a black scoter with its shorter, thicker neck. Both tuber and holder concentrate on watching, and usually the long tube slides down into the stomach so that the tuber can press the plunger of the syringe to force the slurry into the bird, then smoothly withdraw the tube.

Sometimes, of course, the bird will struggle, twist its head and aim its beak at the nearest target, which is either of you, or squawk and squirm until the tube comes out before it can be properly inserted. If there's a small coughing sound, it can indicate the tube is into the windpipe. The tuber then must back off and try again.

One more thing needs attention before the bird is released into the clean side of the pen. That's the feet, especially among birds that have been captive for several days. These are seabirds, and their feet dry out if they're long away from immersion in water. The webbing between the toes is particularly subject to drying and cracking, even separating. So feeders routinely carry petroleum jelly (there's irony for you) to apply to dry skin or foot webbing. That's why we carried a small paper cup taped to the front of our plastic jackets, to carry a glop of ointment during each feeding tour.

We discovered one murre missing a toe from the outermost joint, and the veterinarian on duty said it was probably an old injury, since it appeared to have healed. A diving bird dependent on catching fish for its livelihood might be hampered by an incomplete toe and web on one foot, but this murre was otherwise in good shape.

Meanwhile, the holder's arms get uncomfortably stiff from trying to keep the bird both under control and in proper position for the tuber's ministrations. This can be especially tiring if one member of the team is considerably taller or shorter than the other. That stiffness will surely develop as you work on bird after bird, pen after pen, until all have been fed. The first evening when I was holding birds for Mary Gropp took two and a half hours before the last of 100 or more murres and scoters had been fed. Maybe I was slow. But at the next feeding the team leader, Laura, decided it would be okay to give me a tube and jug to work as a feeder. Mary had been a good teacher, and I hope all of the hundreds of birds I later tubed survived.

One of the U.S. Fish & Wildlife agents, Jim Michaels, later was tagging a bunch of grebes that also were due a feeding, and chose Steve Gropp and me to tube for him. Michaels, who could band and record the numbers faster than the two of us could keep up with the feeding, told us that eight birds had been found dead in the pens the previous night. He had autopsied cach of them, and all had expired because of fluid (slurry) in their lungs.

Mention of Steve Gropp, a serious young fellow wearing dark olive drab coveralls who worked the pens in a quietly efficient manner, recalls the necessity to keep stress for these wild birds at the minimum level possible. That problem is especially noticeable when working with grebes, those red-eyed birds with snake-like necks. When grebes get excited, they squawk in a noisy chorus that is picked up by the birds in the next pen until the raucous cacophony would bring a quiet caution from Steve. It's better to move slower and take a bit more time than to cause the birds unnecessary stress. They've had more than enough.

The time it takes to feed all the birds depends both on how many there are and the number of tuber/holder teams available. At first the birds at Ocean Shores were being fed four times daily, then it was reduced to three feedings per day as the birds recovered somewhat from their ordeal of oil and surf and cold. Later, as more birds were cleaned and fewer dirty birds remained, a feeding could be completed in perhaps an hour and a half.

A second feeding method also was employed at Ocean Shores, but was not uniformly effective. A team of feeders made the rounds of the pens, offering smelt or herring thawed from frozen blocks provided by commercial bait companies. Occasionally a bird would grab a smelt and swallow it in normal fashion, except that of course their ordinary fish diet is alive and wriggling.

Murres, grebes and also loons normally feed by diving to chase small fish underwater. Usually they'll surface holding the fish in their beak crosswise, then swallow it with a deft flick of the head. Scoters, by contrast, feed not on fish but on mollusks such as clams and crustaceans like shrimp. From our human standpoint, the curiosity is that they eat them in the shell, later regurgitating the indigestible portion. Since they are bottom feeders, the scoters stay mostly in shallower water than the fish eaters, and often feed in or near the surf just off the ocean beaches.

The three species of scoter brought to Ocean Shores from the *Nestucca* oil spill, surf scoter, black scoter and white-winged scoter, could all be penned together. Most of them, incidentally, were white-winged scoters. An interesting aspect of handling them was their docility while being fed, although they could be difficult to catch. Dr. Vance Yung of Olympia, a retired state physician and friend of mine who was able to spend several days each week as a volunteer at Ocean Shores during January, had a small male scoter drop off to sleep in the handler's arms while Vance was ministering to its feet. Placed back in the pen, the bird was so quiet that Vance feared it might have something wrong, so he poked it, and the bird awoke, startled out of its nap.

Over on the "clean bird" side of the auditorium the same feeding process went on, employing other tuber/holder teams, with the difference that those plywood pens had netting floors raised above the concrete floor of the auditorium. Newspapers were laid flat beneath them to catch droppings, and bedsheets were spread over the top of the pens so the birds would not be further stressed by seeing workers pass by.

Having been in custody for a week and longer, those birds had experienced also the most stressful process of all, an hour or more of scrubbing in warm water and Dawn detergent (one percent solution) in the washing room. That place was a story in itself, an exercise in organized confusion and splashing: constant wetness, runners carrying plastic pans of fresh water to teams of bird washers working intently on individual birds dunked neck deep in suds and being scrubbed feather by feather to get the oil off.

Washers employ cotton swabs, toothbrushes, waterpiks and gentle but firm probing fingers to work away every trace of that brown, polluting oil from the feathers and from around the beak and eyes. The birds get oil on their beaks from preening, and if they get much of it into their digestive tracts it will kill them. One wildlife biologist was quoted relative to sea otters killed by ingested oil in Alaska: "It turned their livers to mush."

In the bird washroom at Ocean Shores, it could require six or more pans of clean soapy water and repeated rinsings to do the job thoroughly. Immersing a bird for an hour while it is scrubbed clean of oil would be fatal unless the water is warm enough to keep the bird at or near its normal temperature. This is doubly important for birds that already have survived immersion in the ocean's chill after being oiled, have made it ashore, then waited on the beach until found and put in a box or plastic bag for hours before finally arriving in the warmth of the rescue center. That's trauma enough, already.

Between shifts as a dirty-bird feeder over several days of volunteering, I was able once to visit the bird washing room to take photographs. It was a special challenge to try for quality pictures without using flash under less-than-ideal light conditions. Professional news photographers overcame difficult problems with fast film and corrective filters; as one less skilled I can only envy their success. Having to work without flash was again an instance of trying to prevent stress for the birds.

There seemed to be special stress in the bird washing area, not just for the birds but also for the workers. One's emotions need not be dangling from your sleeve to be snagged, at some unanticipated moment, by the plight of a wild thing fighting to overcome something it can't understand. The experience of the bird washing room left some with moist eyes in addition to the probability of wet feet and splashed shirt and pants.

Before providing you some of the experiences and impressions of others who participated in aspects of the bird rescue and rehabilitation operation, some further words are in order about the volunteers, without whom the entire operation could not have been successful. Out in back of the center, for example, was a big garbage dumpster that each day was filled to overflowing, and sometimes needed to be compacted by volunteers in order to avoid the necessity for a costly additional pickup. I was told that on at least one occasion a doctor and a dentist volunteered for the unglamorous and unskilled job of stomping garbage, an example of the spirit of volunteers involved.

One evening I found a volunteer sweeping up in the restroom. "I'm a retired Army sergeant," he said, "and I know how to do this." There were college students happily squishing newspaper, although hoping for early assignment to more interesting work. My wife Mary worked one day crushing newsprint along with a physician from a Seattle health maintenance group and her nurse assistant while they hoped for more suitable work. Mary also worked with other volunteers in the people-food kitchen (as distinguished from the bird-food kitchen where slurry was prepared), and made untold hundreds of sandwiches for hungry workers.

A sometimes boatbuilder from Seattle, unemployed at the moment, was financed by his job-holding neighbors so he could go to Ocean Shores as a volunteer worker. There was also the retired school superintendent who worked with me in the bird feeding operation. That gentleman caught and patiently held those dirty, oil-soaked birds for tube feeding during a long shift in the pens. And there were ever so many more people, such as the lady at the sign-up desk who told me she and her husband spend the winter at Ocean Shores, but enjoy their summers in a cabin at Eagle, a small community on the upper Yukon River in Alaska just west of the Canadian border of Yukon Territory.

Among the bird washers were Emma from Sydney, Australia, and Reggie from Hoquiam, just a few miles east of Ocean Shores on Grays Harbor; Bruce from Guemes Island and Sandy from Tiger Mountain, near Issaquah east of Seattle, and Michelle from Bellingham; Kathy from Montesano and Dawn from Oak Harbor--the list included more than 2,500 names recorded at the Ocean Shores center alone. There were scores of other workers at such places as Neah Bay near Cape Flattery on Washington's northwest corner, and the folks who patrolled beaches seeking oiled victims of the spill, and more who cared for and cleaned birds at Friday Harbor on San Juan Island.

The response of volunteers on the remote west coast of Vancouver Island in British Columbia was unique in a special way. It was the volunteers, including Native people and their tribal leaders, who put pressure on provincial and federal authorities of Canada to force increased attention to the serious oil pollution on those relatively pristine shores.

There were also the supervising people from the Washington State Departments of Ecology, Wildlife and Emergency Services, plus the Grays Harbor Sheriff and his staff, the County Department of Health and County Emergency Services. The participation and cooperation of both the municipal authorities and residents of Ocean Shores was exceptional, and continued for six weeks. The U.S. Fish and Wildlife Service people, led by Don Kane from its Olympia district office, and members of the U.S. Coast Guard also provided help as part of their supervisory duties. Whether volunteers or people in jobs that required a response, in most instances those involved gave more time and effort than anybody could ask or expect.

Sometimes disagreements and frustrations occurred, and we'll pay them some attention. No such operation, organized in response to an emergency situation, is going to be perfect, and this was typical.

In contrast to the response of some other companies in similar situations was the part taken by Sause Brothers Ocean Towing Company of Coos Bay, Oregon, owner of the barge *Nestucca* and tug *Ocean Service*. Company officials were up front in taking responsiblity for the oil spill, and paid without stint the expenses for beach cleaning and bird rehabilitation. Dick Lauer, the company's manager of bulk shipping, spent the last of December, most of January and early February facilitating work and paying for cleanup operations along the coast. The towing company had insurance, of course; but the impressive aspect was the attitude taken by its officers: complete cooperation rather than buck passing and denial. Working through a subcontractor in British Columbia, however, the Sause response was not as well received.

"We're part of the community, too," was how Rick Kimberley expressed it to me in a phone conversation at Sause Brothers headquarters in Coos Bay. "We plan to stay in business and to continue as good neighbors."

Perhaps not too many people are aware that birds can indeed display emotion, but individuals who visited the bird release area near the Ocean Shores fire station in late January could see avian happiness demonstrated. My friend Vance Yung told of watching the obvious elation shown by clean and rehabilitated birds released into the freshwater canal flowing southward toward eventual discharge into Grays Harbor. "They were just so happy to be free. I saw them swim sideways, roll over in the water, even swim upside down. They were having a ball. And remember how they'd peck when we handled them inside, yet where the water widens into sort of a pond, it seemed they'd lost some of their fear. If you stood by the bank and raised your hand, these wild murres and grebes and scoters would swim over to see if you had a fish to feed them. They had lost some of their fear of us, and I just hope they return to normal wildness. When I'm out sailing (on Puget Sound near Olympia), it's unusual to get closer than 100 yards to a grebe or loon."

Maybe that's true for Vance in his craft with its tall sails, but our experience in a canoe is that we can paddle as close as 50 yards before a grebe will dive. That's close enough.

Now if those birds could just learn to take off and fly upwind at the first whiff of oil on the water!

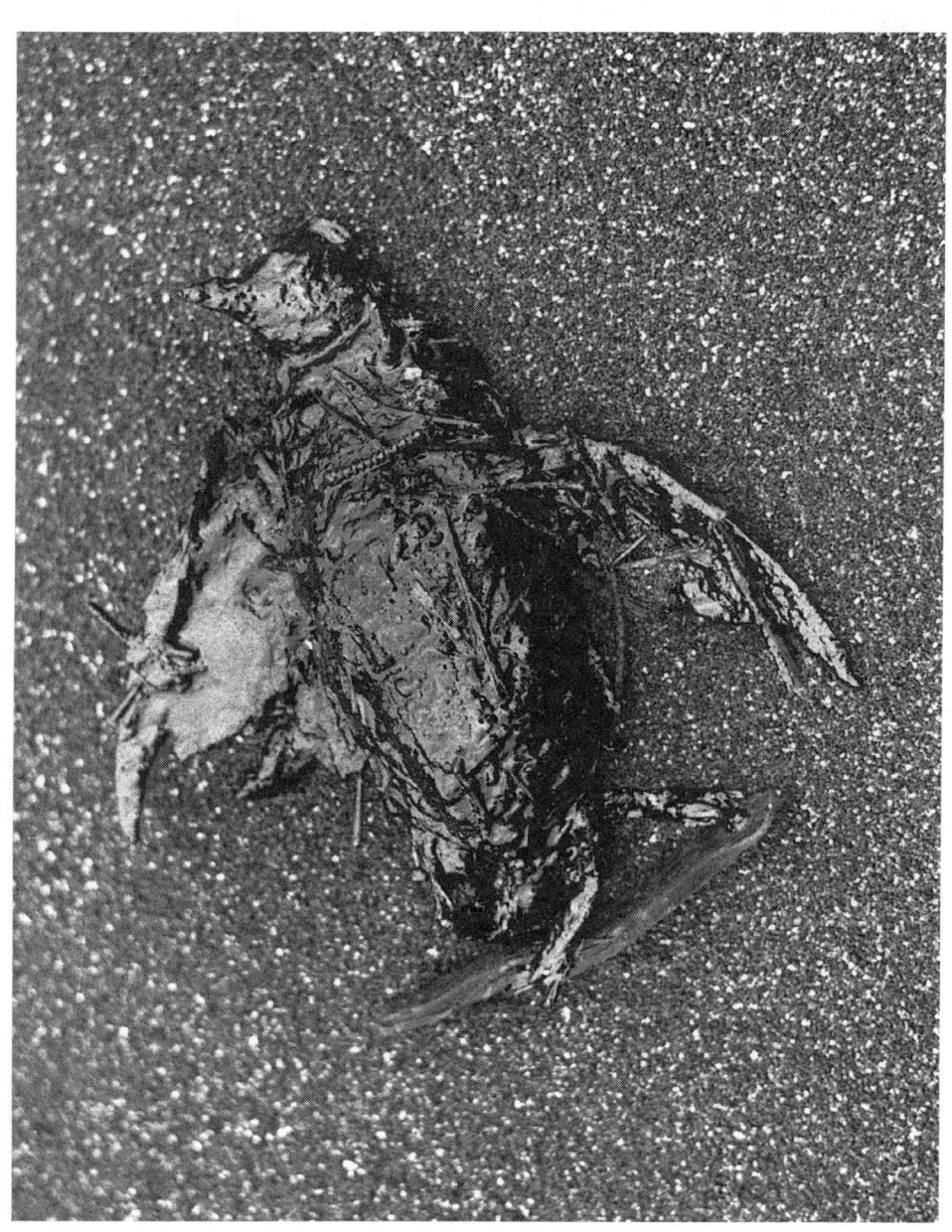

Termed a "dipper" by oil cleanup workers for the obvious reason that it appears to have been literally dipped in heavy black oil, this bird may be an auklet, one among many thousand avian victims of the *Nestucca* spill. (Photo by Larry Workman, QIN)

Cape Flattery,
northwestern-most
of the south 48 states,
stands tall against
Pacific storms in this
unusual view,
photographed by
Ruth Stennett.

North of Taholah, Lewey Kittle (left), on-scene coordinator for the Washington State Dept. of Ecology, and his assistant, Jim Oberlander, spill response team leader, inspect a stretch of beach impacted by bunker fuel oil spilled from the barge *Nestucca*. In foreground are clumps of oil shiny with reflected sky light. (Photo by Larry Workman, QIN)

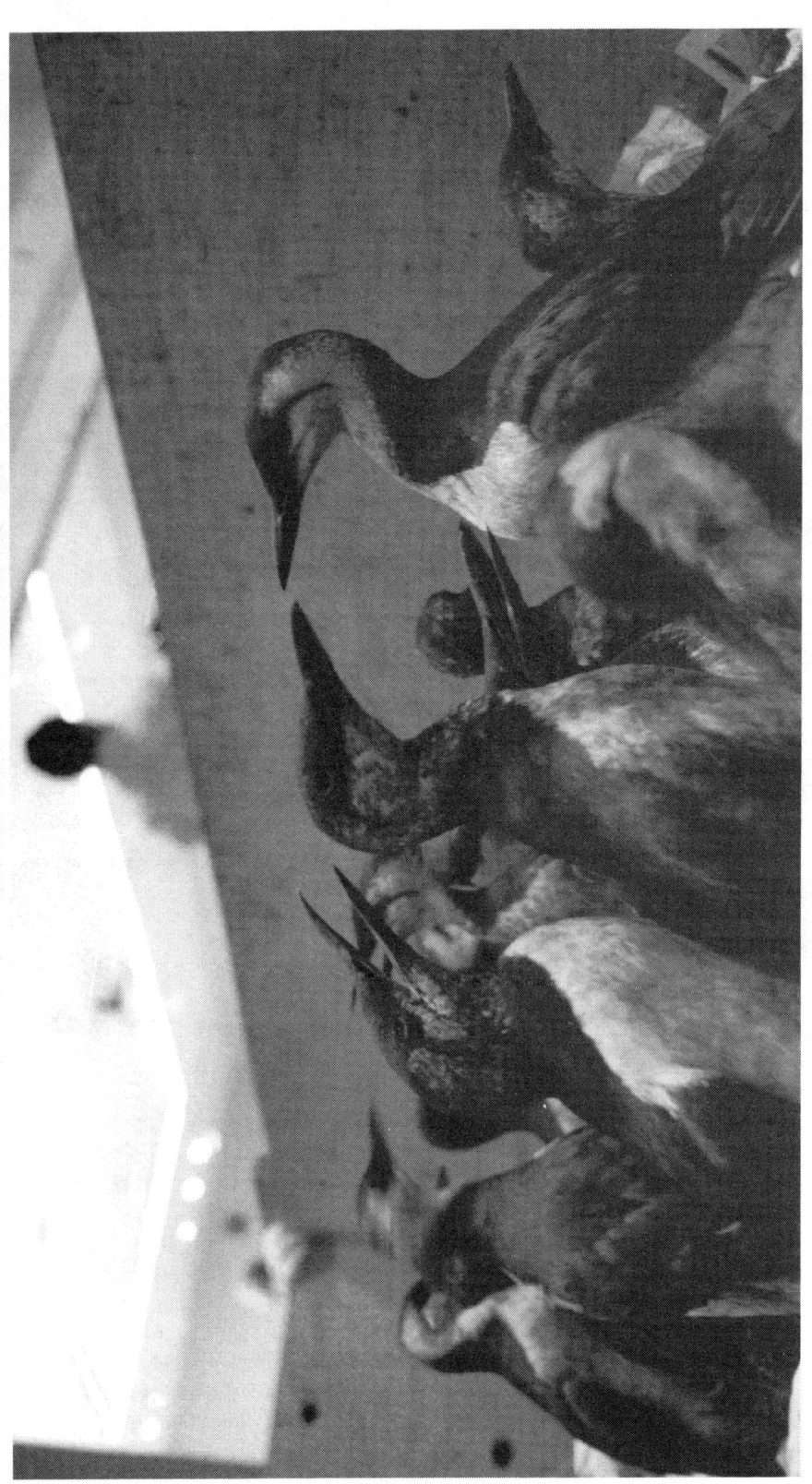

Looking up at a gaggle of oiled murres inside a plywood pen in the "world's largest bird hospital." Ocean Shores Convention Center was converted into a rehabilitation facility after the *Nestucca* oil spill. These murres, all captured the same day, were penned together waiting their turn for washing. (Photo by Larry Workman, QIN)

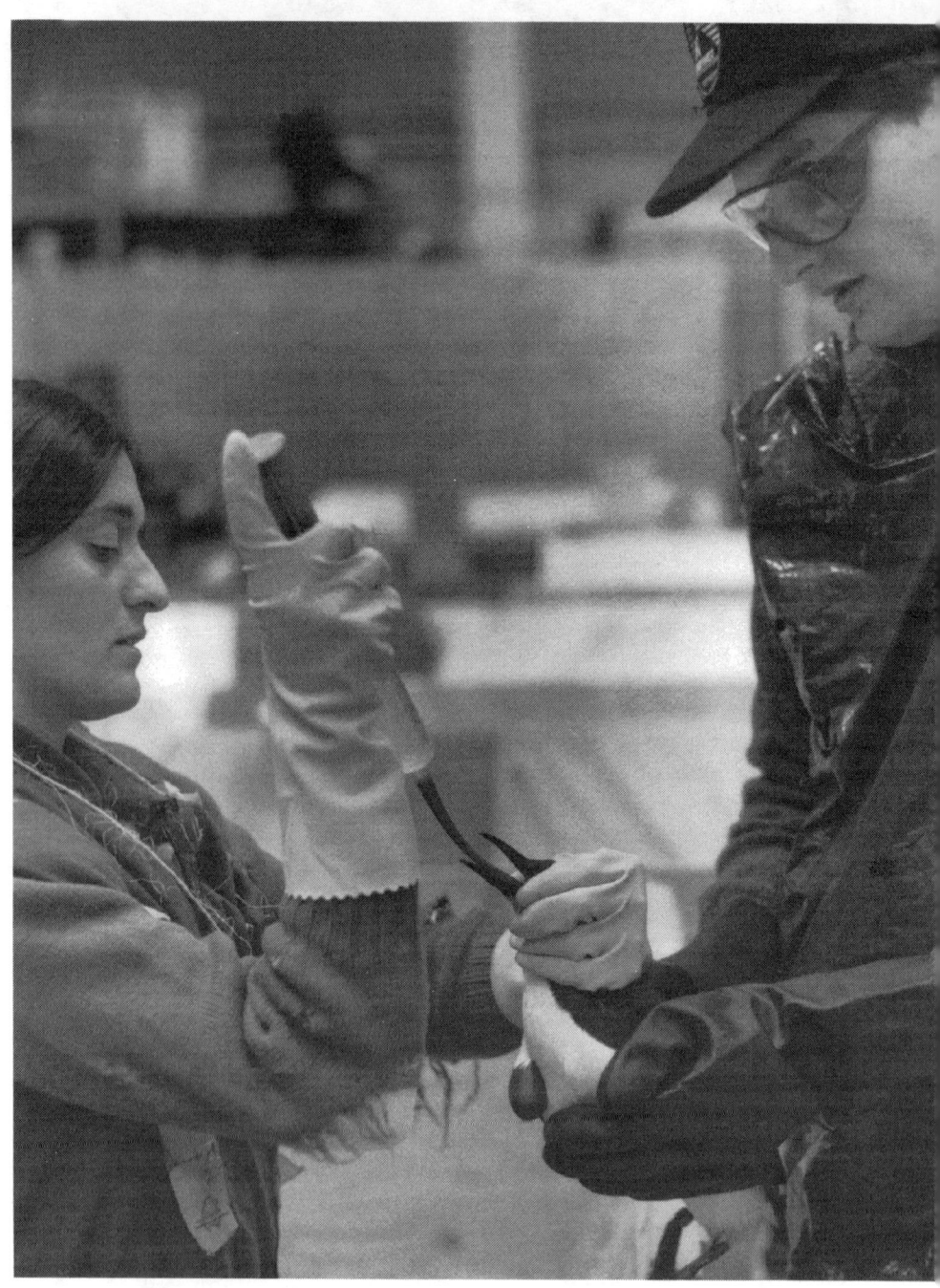

"Tuber" on the left, "holder" on the right wearing safety glasses, both of them unidentified, demonstrate how the catheter is slid down the gullet of a murre while the tuber holds her left hand over the bird's head and eyes, keeping the beak open. It appears the tube is well down into the murre's stomach, and the tuber is about to inject "dinner," a nutritious slurry from the 60 cc. syringe. (Photo by Larry Workman, QIN)

Two bird washers concentrate on murre immersed in warm water and Dawn liquid suds as they cleanse its feathers of oil in the Ocean Shores bird rehabilitation center during January, 1989. Bird washers found their shifts exhausting, both from the long hours of standing work in the wet steamy atmosphere, and from the concentrated drama of the effort to save oiled birds in their care. (Photo by Steve Vento, Olympia, provided by Wash. State Dept. of Ecology)

Lone jogger paces tideline on Vancouver Island's Long Beach in Pacific Rim National Park, as January surf blasts offshore rock outcroppings. This 20-mile beach was oiled when widely dispersed bunker fuel from the barge *Nestucca* was reconstituted by on-shore winds. It created a mess to which residents and tribal folk responded immediately, but the agencies of government less quickly. Volunteers said the official cleanup bypassed these beaches. (Photo by David Webster)

Sause barge *Nestucca*, her chain bridle still dragging the broken cable, waits inside the Columbia River bar. The hull gash, made visible by the ends of wooden wedges hammered in to stop oil, is on the starboard side in line with the deckhouse. Note bow shield behind which *Ocean Service* crewmen sheltered when put aboard to pass a line to the tug. As huge swells broke over them, they retreated inside the deckhouse until morning. (Washington State Dept. of Ecology photo)

Towing vessel *Ocean Service* in early 1994 after a voyage to Hawaii, is moored at the Sause Brothers Ocean Towing Company's operational headquarters in downtown Coos Bay, Oregon. The 120-foot tug, powered by twin 1,900-horsepower diesel engines and rated at 193 gross tons, was built by Burton Shipyard about 1966. She hauled petroleum barges on the East Coast for some ten years prior to acquisition by Sause in 1984. (Photo by Sause Brothers Ocean Towing Co., Inc.)

Shipyard workers in Portland dismantle towing gear on aft deck of the *Ocean Service* following the oil spill off Grays Harbor. A curved wear plate on the deck next to a pair of red fuel tanks appears to have been removed from the level-wind assembly just forward of two workmen. Hardhat of a third worker is visible over framework next to the main wench drum, emptied of its wire rope remnant after the break. (Photo by Capt. G. Kurt Greiner, USCG (Ret.), for Washington State Dept. of Ecology)

Washington Conservation Corps crews were among the first to respond when oil from the *Nestucca* blackened Ocean Shores beach north of the Grays Harbor entrance. This crew of five took a photo break, behind them the ragged dark smear of bunker-C fuel oil left by receding surf. The crew chief at left was Will Williamson, according to Steve Sherlock of Wash. State Fish & Wildlife. Kirk Thomas of State Ecology said more than 70 WCC people put in 15,000 hours cleaning beaches and recovering oiled birds, live and dead. (Washington State Ecology photo)

Eagle Air helicopter lifts oiled debris from beach for flight to a dump station where it can be loaded onto a truck for transport. Considerable oiled debris was burned, while some was buried in landfills. This may have been photographed in the Grays Harbor area. Typical helicopter load weighed from 600 to 800 pounds, depending on the weight of fuel in the aircraft's tank. (Photo courtesy Washington State Dept. of Ecology)

Jody Drew from Seattle, bundled against the chill ocean wind of January, 1989, at Ocean Shores, observes a recently-washed group of murres in a tank of water fresh from a municipal hydrant. The tank was from State Dept. of Natural Resources forest fire fighting equipment. "I'm watching for sinkers," she explained. "Sometimes one of them can't float. My job is to save it from drowning if it sinks and gets stuck by the current against the outlet screen." (Photo by David Webster)

Petroleum-based horror in a different form visited Washington State when a 16-inch diameter pipeline burst on June 10, 1999, sending 277,200 gallons of gasoline flooding down Whatcom Creek toward downtown Bellingham. As it flowed through the city park, two 10-year old boys, Wade King and Stephen Tsiorvas, lit a match and touched off a massive and fatal fire. Liam Wood, 18, fishing downstream, also died. This view of the holocaust was made from downtown. Investigation was continuing at the time of this book's publication, as were efforts to increase local and state control of federally-regulated fuel pipelines. (Photo courtesy Whatcom County)

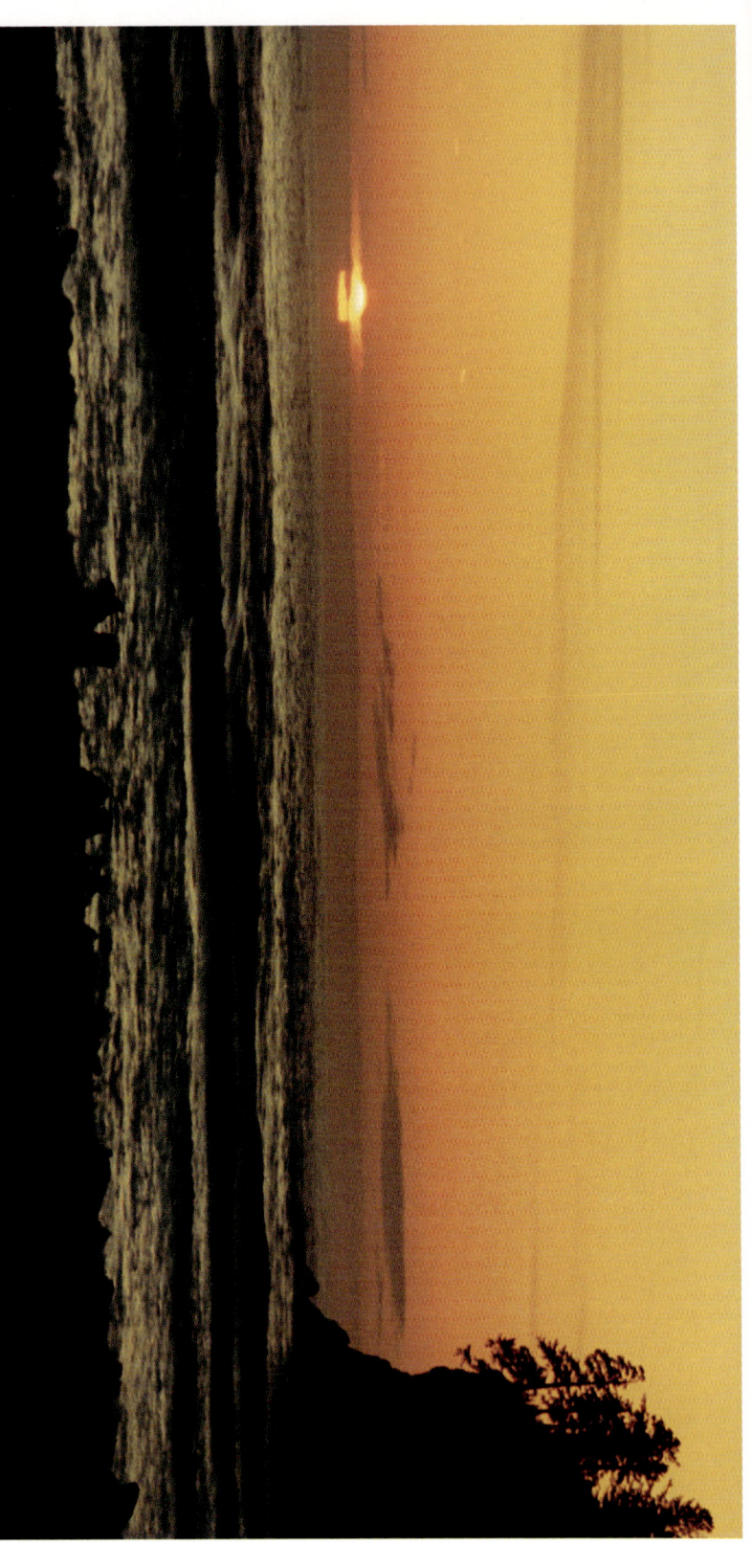

Surf rolls in under a setting sun at LaPush, home of the Quillayute tribe and the river of the same name. James Island is at right. Fishing boats slip in and out of the river mouth just this side of it. In earlier times, a fortress was maintained atop the island. The Quillayute people wintered in traditional split cedar plank homes near the river mouth, but in season moved upriver or among coastal fishing, hunting and other food gathering sites. The Olympic Peninsula provides a succession of such memorable scenes. (Photo by David Webster)

8

Personal and Statistical

ONE ON ONE
by Constance Perenyi

What I remember most is the smell. Even now one whiff of Dawn dishwashing detergent and I am no longer standing at the kitchen sink but in a...gymnasium filled with frightened animals (8-1). I have volunteered to care for birds after three major oil spills off the coast of Washington. And while I am a veteran of sorts, I still have no idea who discovered the multipurpose utility of Dawn. I do know it is the best solvent to remove oil from feathers and fur. Pumped by the gallon, this pungent blue liquid is vital to rescue efforts. It can also trigger memories.

I began as an artist, trained in visual composition and steeped in love of nature. For years, my work was abstract, the birds and other animals implied but obscure. Dissatisfied,...I began to fill the gaps with practical experience by working for Seattle's Woodland Park Zoo..., (learning) about birds by preparing their food and cleaning their enclosures. I read everything I could find about birds, absorbing facts and experiences and eventually translating them visually. But it wasn't until I worked at an emergency oil spill clinic that I stopped observing animals at a distance. Surrounded by injured birds, I could not separate myself from their vulnerability. I was sickened by the oil on their feathers, evidence of human insensitivity to the environment. At the same time I was heartened by the humane concern of my fellow volunteers.

(After the Christmas, 1988, Nestucca spill)...inadequate supplies of hot water at a temporary clinic hampered washing efforts. I joined other volunteers in an attempt to keep the unwashed birds fed and hydrated until we could move to better facilities. Exhausted after my first day..., I checked into a motel, tried to rest, and awoke abruptly. I looked around...and thought I saw shadows of murres and guillemots on the floor. They broke in dark waves against the bed, vanishing every time I reached out. I forced myself to go back to sleep, and then dreamed about a small white-winged scoter I had held during the day. Unlike the other birds, she was quiet. I stared at the soiled feathers on her back and sheltered her head with my hand. As I rocked her, her heartbeat weakened, and I repeated, "I'm sorry, I'm so sorry." For weeks afterward, I fought to rid myself of this nightmare and often awoke in tears.

Time has passed, the dreams have faded, but not the scars on my hands. These commemorate the feisty energy of the murres and my often unsuccessful attempts to catch them without getting caught first. Jousting with one after another, I learned to respect their tenacity.

At Port Angeles (in 1985 following the *Arco Anchorage* spill),...the washroom coordinator handed me a double-crested cormorant that had been washed and generally de-oiled. My task was...to clean each feather on its head with a WaterPik. For hours I sat on a folding chair with the bird balanced between my legs and the Pik humming...next to us. With one hand, I supported the cormorant's neck. With the other, I focused the water and combed across the feathers in short, methodical strokes.

I began...cautiously, washing the bird while it noted my every move. Earlier, a veterinary technician had closed its sharp beak with tape so the bird could not bite or spear its handlers. But when my charge began to blow bubbles through its elongated nostrils, I removed the band. The cormorant instantly relaxed and seemed more curious than dangerous.

...This was an extraordinary opportunity. I had handled captive birds at the zoo and wild ones at other clinics, but I had never examined one so closely.... As I sensed the cormorant's damp heat radiating through my clothes, I experienced warm-bloodedness in a new way. The bird and I exchanged body heat everywhere we touched. I could feel impressions of webbed feet on my legs and feather marks along the inside of my arms.

Unlike others around us, this bird appeared healthy and unafraid. As it looked around the room, the cormorant seemed to watch the day unfold as a spectacle rather than a trauma. It was calm and seemed willing to cooperate, which enabled me to consider every feather tract I cleaned. The more I contemplated the depth of its blackness, the more detail I perceived. Points of turquoise outlined pale green eyes. Burnt-orange skin marked the bird's throat in colorful, featherless contrast. Against my skin, its snakey neck and armored feet felt reptilian. I could even feel its small, flexible gular pouch.

As the afternoon unfolded, the bird became my sole focus.... I became oblivious to the hectic activity around us,...and soon it seemed as if the bird and I were the only beings in the room. ...None of us could reverse the damage (of the oil spill), but... by working on the problem at its most elemental level, I had found a way to make amends with a single bird. ...For five hours I...enjoyed the company of another creature and...almost regretted ending our time together. Reluctantly, I finished the task and lowered the cormorant into a pool with other birds. This was the final test; if it could float, its feathers were oil-free. Success. The bird swam off and the washroom volunteers cheered. The cormorant looked strong, and after it regained energy and natural oils, other volunteers would release it on a clean shoreline miles from where it had been found.

But release was weeks away, and as I watched the bird, I realized I would never know what happened to it. The mortality rate of animals rescued from an oil spill is depressingly high. Even (after)...cleaning, many later succumb to lethal doses of ingested oil. And of those released, few remain capable of reproduction. I knew the statistics, but I...still like to think that the cormorant is alive, a survivor. If a bird can be said to express a will to live, this one seemed to do so.

It is the hope of saving even a few animals that motivates volunteers at an oil spill clinic. Confronted by death, we work hard to preserve life. We do what we can, and in the process are changed.... I realize my efforts with the birds have helped me more than I can ever hope to help the birds.

* * *

NO BETTER PLACE TO BE

Ranger/naturalist Paul Crawford of Olympic National Park has worked through three major oil spills affecting wildlife along its shores.

Although the earliest spill, that of the *Arco Anchorage* in Port Angeles harbor, did not touch the park boundary, it involved birds from park waters and occurred within a couple of miles of park headquarters.

Crawford recalls working in the Port Angeles bird rescue operation after that pre-Christmas spill. He supposed that all volunteers would leave on Christmas Eve, but to his surprise the reverse happened. More volunteers showed up.

"I was working on a bufflehead, I think, (cleaning oil from the duck's feathers) and I asked a woman volunteer where she was from. She said Seattle. She and her husband had attended the Christmas Eve service at a church, and as they left they wondered what to do next. 'Let's go to Port Angeles to help. I can't think of a better place to be on Christmas Eve'."

The park ranger added: "Once you have worked with an oiled bird, you would understand. It's a spiritual thing."

* * *

GOOD FROM DISASTER

It seems hard to believe that something good could come out of an oil spill, which seems to be unmitigated disaster. But perhaps if we support the recommendations from the state, tribal and regional committees and agencies working to prepare, equip and train both workers and volunteers to better cope with situations following oil spills.... Of course the best way is to prevent future spills, but it is necessary to have plans and skills and equipment ready when one does happen.

Who is responsible? I am responsible, and so is each of us.

* * *

SQUAWKS

Returning to the birds cared for at the Grays Harbor Bird Rehabilitation Center from December to February, first in Hoquiam and then at Ocean Shores, all was not sweetness and light. There were "squawks" over some aspects of the bird rescue work, as reported by Theresa Willeford in the weekly *Willapa Harbor Herald* (8-2).

"Some local volunteers involved with the oiled bird cleanup have had their feathers ruffled by what they feel is mismanagement at the Grays Harbor operation.

"Patty Hodel, a 4-H leader from Raymond, along with six other volunteers from the area, attended one day of training Jan. 3 at the Ocean Shores Convention Center. The birds are being washed there by hundreds of volunteers under the supervision of Alice Berkner, head of the Bird Rescue Research Center in Berkeley, California....

"Both Mrs. Hodel and volunteer Kay Fitzgerald said they were not impressed by what they saw at the center, with inefficiency leading to unnecessary deaths among the birds and a slower pace of cleaning than necessary.

"Ms. Fitzgerald said it appeared that 'nobody was in charge' at the facility, while Mrs. Hodel said the volunteers were required to wash the birds in water that was 108 degrees Fahrenheit. The birds would start panting from the heat,' she said. 'And they told us when this happened to put them in warm water so they won't stress out. Well, all that heat just opens the blood vessels and the blood pressure drops, stressing them out further. That water was so hot our skin was beet red.'

"Also, she said, the center requires two persons to wash and rinse each bird, one to hold the animal, and another to do the scrubbing and rinsing. The crew members frequently left the faucets on when they were not in use, she said, with water pouring out and a continual shortage of hot water.

"Mrs. Hodel said many dirty birds sit in cages waiting to be cleaned while time is wasted on inefficient procedures. 'Even one washing would help immensely,' she said.

"After the 8:30 a.m. to 5:30 p.m. session, the volunteers were allowed to take 10 dirty birds to Vetter's Veterinary Clinic to care for, because Dr. Don Vetter is a licensed wildlife rehabilitator. Mrs. Hodel said the understanding was that...officials would let them know when the birds could be released, and also replace them with more birds. One of the 10 birds died in transportation. 10 to 15 volunteers from the (Raymond) area took shifts cleaning and caring for the nine survivors. Vetter supplied the facility and supplemented the supplies provided by the center.

"At Vetter's, the birds were washed in lukewarm water and a special cleaning solution, DSS (dioctyl sodium solfosuccinate). They were fed ground frozen salmon, a nutritious liquid called electrolite pedialyte, and herring. Grays Harbor officials have ruled that no one under (age) 18 can help clean the birds, so local boy scouts and members of the Holkum Yolkum 4-H Club helped by crumpling paper for the cages and cleaning the bird site. Both Mrs. Hodel and Ms. Fitzgerald said they did not understand why children can pick up birds on the beach but not participate in the cleaning.

"Mrs. Hodel said it took only one person to clean each bird at the clinic. She...had experience in cleaning chickens for bird shows, and...the cleaning techniques were 'just common sense.' After cleaning, the birds were moved from cages to a...moderately heated and ventilated pen. On Friday, Mrs. Hodel said the birds were clean, healthy and ready to be released. She and the other volunteers were waiting for word...on when the birds could be released and...be replaced with more birds. She...could not see why the birds could not be released in...Willapa Harbor, where no oil...has been reported, but that this was ruled out....

"The volunteers' hopes of seeing the birds released were thwarted on Saturday, when bird experts from Grays Harbor swooped down on the veterinary clinic and confiscated the birds. 'We feel like we've been slapped in the face for all our efforts,' Mrs. Hodel said. 'They took the birds without calling. There was no 'thank you' or remarks on what a good job we did. We won't even get to see them released. It's upsetting after all the hours we put in over there training at Grays Harbor.'

Don Kane, U.S. Fish & Wildlife Service field response coordinator, said the birds were returned to Ocean Shores...for rewashing because...(their) feathers were wet, and one...had a small amount of oil residue. 'DSS is a wetting agent which might explain why the birds were wet,' he said. 'We have found through extensive studies that Dawn dishwashing liquid is one of the most effective products for removing oil from...birds. When the bird has been cleaned, the feathers should come up dry and fluffy, not wet. This is because the feathers are insulation to keep the bird warm and dry. Also, we cannot have any oil on them when we release them.'

"In defending the facility techniques, Kane pointed out that Ms. Berkner has years of experience in cleaning oiled birds and is regarded as a national authority on the subject. He also said he thought the 108-degree temperature sounded 'a little high,' and that it more likely to have been between 98 and 104 degrees, which he said is not all that hot. On the matter of two persons handling birds, he said that he could not see how a person could safely handle or clean a bird alone. He said the animals are strong, frightened and have powerful beaks.

"'As for the thank you,' he said, 'I can't say they were or they were not thanked, (but) I can assure them that we sincerely appreciate the efforts of every volunteer. If they were not thanked, I apologize, and I can call them and personally thank them. We have more than 500 volunteers going through the facility (daily) and we do our best to thank them all, but I'm sure we do miss a few.... Kane said no birds were being released in Willapa Harbor because experts must first determine all the oil impacts on the coast (and at that time had not yet certified that Willapa Harbor was oil-free)."

Another who had objections to the bird cleaning operation was Hilary Richrod of Aberdeen, an experienced rehabilitator who asked, among other questions, "Why not wash birds 24 hours (a day), if the facility has the capacity? Keep the oiled-bird pens dark, remove them one at a time, and wash, wash, wash. Berkner said they would, and then changed her mind, even when continuous hot water was available.

"There are rather weak arguments for not washing birds continuously; one is that they need a 'rest period.' A dandy thing for healthy birds, but birds covered with poison? Birds do need a rest and to be warmed and fed in the first few hours, but to make them 'sleep' overnight is not necessary.... I don't know why the birds weren't given vitamin B1...or codliver oil to help correct and remove the petroleum in their systems. These are standard rehabilitation techniques for seabirds. There were certainly plenty of people available..., plenty of money, plenty of everything but organization. ...The washing was incredibly slow, only 100 birds a day.

"I have taken care of...murres, and they are gentle, social,...mildly curious, 'upright' little birds. Seeing them dying...for lack of care was like...(watching) children in the same condition. They are long-lived birds that reproduce slowly, and we have damaged them very badly. They don't even compete with us for the fish they eat, or the place they live," Ms. Richrod said.

With that much of the negative comment, leaving out even more provocative and scathing remarks attributed to other persons, we'll turn to Alice Berkner herself, interviewed at her Vancouver, Washington, apartment in 1993.

Responding first to the comments by Ms. Richrod, Berkner said "Hilary I think had been doing some bird rehabilitation on a very small scale, and was trying to get a certificate from Wildlife. (On a broader scale)...I think there was a bit of territorial resentment among some of the local rehabbers."

Webster: "They didn't like...an outside expert coming in?"

Berkner: "Yes, indeed. But it's always hard. I try to work people into the system, but when they become disruptive, that's different. And when I've done thousands of birds to their one or two, I feel that I do indeed have more knowledge.

"I remember her (Richrod) saying we should work all night washing birds, and we used to do that years ago; but our research showed that the birds...cleaned at night have a much higher mortality rate. It just made a terrific rise in mortality of all the birds cleaned at night. Well, we're all warm-blooded critters, and that's the reason why (in a hospital) elective surgery is never done at 4 o'clock in the morning. The mortality tends to spike. Birds are susceptible to this, too."

Webster: "Is it fair to say birds operate on a very different level than humans?"

Berkner: "They do and they don't. Humans are very susceptible to light."

Webster: "Well, I was thinking of birds as very programmed."

Berkner: "Yes, instinctive. And you have to be aware of that. I think people try to make the birds fit their idea of how things should be done, rather than (to) study the animals. You try to put yourself in the bird's position."

Webster: "And we don't understand birds."

Berkner: "No, no, we don't. Or at least if we watched them in the wild, we would get an idea of what they are used to. Herons and cranes, for instance, you want to give them something to hide behind. Pelicans like to perch, so we need to provide them a perching place.

"I'm scientifically trained, (and lived with) a physician (her father) and physicist (her former husband) for long periods of my life. Believe me, you learn to look at things very logically, and try not to let emotions (rule your actions), because if you do,...you'll be...asked 'Well, why did you make this decision?'

"You're working (in rehabiliting oiled birds) with U.S. Fish & Wildlife Service biologists, and if I don't want to go along..., I'd better have a darn good reason. It's unfortunate there were bad feelings. I'd like it to be great, but unfortunately there are...personalities that just don't mesh. I always try. It wouldn't be just, wouldn't be fair not to."

Turning to another aspect of the *Nestucca* experience, Berkner recalled her arrival on Grays Harbor during the Christmas vacation to find birds being cared for in a school. "Here we go again," she said was her thought at the time, since she also had supervised bird rescue following the *Arco Anchorage* spill at Port Angeles. "Why have you done this, don't you remember the *Arco Anchorage*? It was Christmas, the same time of year, (and) we had to transfer just about the same number of birds from both facilities (when school reconvened after the holidays). At Port Angeles, we found a warehouse (to move to). The Ocean Shores Convention Center was the better facility compared with Port Angeles, because there was office space and kitchen. But it was like, 'Hey, guys, when are you gonna get the message?'

"If you wanted to see a well-functioning spill (response), ...just smooth as silk, it was (following) *Tenyo Maru*, at St. Edward's."

The former Catholic seminary at Redmond, now a Washington state park, was used as the bird cleaning/rehabilitation center for that 1991 spill, and oiled birds were transported there from the ocean beaches. Jim Oberlander, the spill response team leader for State Ecology, recalled that the Lake Washington site worked out well. "We finally learned," he said, "that the birds can take the transportation" to where most volunteers are available closer to their homes. "What's established now (for future spills in Washington waters) is they'll use it. You have parking, bus service, water and sewer, electricity, extra rooms for training people."

Berkner apologized, saying "I don't like to compare spills, because they are so different. But with the *Exxon Valdez* (in Alaska) coming so rapidly on the heels of the *Nestucca*, I couldn't help but compare. If I ever see Dick Lauer again, I want to tell him, 'Gee, Dick, how many times I wished you people had been responsible for the *Valdez* spill,' because Sause Brothers' Rick Kimberley and Dick Lauer were so good to work with. Absolutely super. I would clean up their dirty birds any time. And Exxon was so horrible.

"Initially when I went up to Valdez I was working with Alyeska, and I...would have preferred to stay with Alyeska rather than go over to Exxon, but unfortunately that wasn't the case. It...just got out of hand up there. It was some of the same things. Volunteers show up, and want to know where they're gonna stay. And it's 'Wait a minute, you don't understand, I don't arrange housing. You're asking a little too much now, with 500 birds that just came in, and 500 volunteers, and the press is showing up, and I've gotta get all these supplies. Really, I'm so glad you showed up, but would you mind taking care of your own housing?"

The population of Valdez, Berkner said, "went from 4,000 to 12,000 within three or four days." She mentioned calls from congressmen in Washington, D.C., seeking arrangements for constituents coming north as volunteers. "The White House called, wanted to know about housing, said they had a volunteer hotline, and we told them 'You don't understand, we don't need volunteers, we have no room for them.' But they said, 'You don't understand, this is the White House.' I said 'You don't understand, this is Alaska. I'm Alice Berkner, and I'm in charge of the bird cleanup. There's one thing you can do for me.' And they said 'What's that?' And I said 'Just keep out of it,' and I hung up.

"People in Alaska wanted to volunteer," Berkner explained. "It's their home that was wrecked, (yet) we had...people calling from Texas, and jeez.... You know, the first two weeks you couldn't call out of town, because...they didn't have enough phone lines. I needed to talk to the guy (running) search and rescue over in Seldovia, and I had to go to the police station in Cordova and have them telex (for me).

"We ended up with four bird rescue centers. The initial center was set up at Prince William Sound Community College, in a dormitory, right in Valdez. Second, we had a small facility in Kodiak, down at the national guard armory. And then there was a small facility at Homer, and we had a building...totally modified...in Seward. The majority of the birds came into Seward, as we figured they probably would, about 1,600 birds in six months, not a lot. Fortunately, the spill happened before the migrants (birds) came in.

"I've been working up there since 1975 (periodically), and learned that you lean very, very carefully on the local people for their expectations. When I first went down to Kodiak, they were in an uproar, because the Exxon man had said, 'We're gonna hire people to clean birds.' They didn't like that idea, didn't want Exxon hiring anybody down there. Why not? 'Well, you don't understand. We want to volunteer.'

"I walked in, and I just interrupted and said, 'Excuse me, but if there are going to be birds cleaned down here, I'm the one who's going to be doing it, and since when did Exxon set my hiring practices? If I hired 40 people off the street, I might get two or three good duck scrubbers. It's something you're born with. It's genetic, it really is. You either have the duck-scrubbing gene or you don't. My daughter does a fantastic job. My sister has the gift. But my son couldn't clean a bird if his life depended on it. Some people can do it and some people can't. It's as simple as that.'

"So I quieted them in Kodiak. The mayor got up and said, 'Oh, if that's the case, everything's fine.' So I turned around, got on the plane again and went home (to Valdez). Later they had a town meeting, and I went down and talked to them about birds, and when we were finished, got one of the nicest compliments I ever had in a spill. Some fellow got up and said 'Thank you very much. I feel like you're an Alaskan speaking for us, for the rest of us.' And I just teared up and said, 'Sniff, thank you very much'."

* * *

BIRDS RELEASED & OTHERWISE

The efforts of the *Nestucca* center and its thousands of volunteers at Hoquiam and Ocean Shores resulted in the eventual release of 1,027 rehabilitated birds, most of them on Puget Sound off the National Wildlife Refuge at the mouth of the Nisqually River east of Olympia. Rehabilitators released a total of 72 birds, the Neah Bay center released 34 and USF&WS the rest. They included 706 murres and 281 scoters, of which there were 7 black, 24 surf and 231 white-winged, plus 19 not specified. Others released were 27 western grebes, one horned grebe and 3 unspecified grebes, one goldeneye, two dunlin, a canvasback, a bufflehead, a rhinocerous auklet, a loon and two puffins.

The center received 7,245 dead birds, 932 died there and 1,133 were euthanized as being in too poor condition to survive. Of the 3,092 oiled birds received alive, the 1,027 that were released represented 33.2 percent survival.

There are no simple answers to questions about how many birds were killed by the *Nestucca* oil and what percentage of the vulnerable population they represented.

To begin with, nobody knew how many birds were out there, and only since *Nestucca* have studies been initiated. Also, since 1989 the breeding population of murres has declined drastically, probably from a variety of causes. Murres and other alcid species of marine diving birds were the main victims of this and most marine oil spills.

Because of seasonal migrations, miserable winter weather, lack of money for research and other factors, little information existed on winter offshore bird populations. Even the spring, summer and fall information on offshore bird numbers was scanty, and knowledge of winter numbers almost non-existent. The Audubon Christmas bird census does not include offshore birds, and few inshore birds are counted because of usually poor visibility from the beach.

Two ornithology students from The Evergreen State College, Olympia, Eric Larsen and Scott Richardson, conducted surveys of oiled shorebirds wintering at Grays Harbor immediately following the *Nestucca* spill.

Richardson and Larsen were early arrivals on the ocean beach that first morning, and helped rescue some of the oiled seabirds coming ashore. They later wrote of a December 27 census of shorebirds along the 11 miles (18 km.) from the harbor entrance northward to Connor Creek with 31 percent oiled (3,574 of 11,708) which dropped to less than 1 percent (213 of 15,000) on 18 January (8-3).

Larsen and Richardson believed that most of the oiled sanderlings, dunlins, black-bellied plovers, semipalmated plovers and western sandpipers probably died as a result, although few were found dead. Such birds, they explained, left their feeding flocks and tended to disappear eventually, although it was unknown whether from predation or by succumbing to the effects of ingesting oil preened from their feathers.

The Larsen/Richardson report was noted but seemingly not taken into account in estimating total mortality of seabirds from the *Nestucca* incident.

Except for a couple of one-day boat excursions, no surveys of offshore bird populations in winter had been done prior to January, 1990 (8-4). That first major survey, conducted January 3-14 from the air, covered the area from Estevan Point, midway of Vancouver Island (at its widest spot), south to the Columbia River boundary between Washington and Oregon. Details concerning the survey are in a report prepared for the Washington Department of Wildlife by Ecological Consulting, Inc., of Portland, Oregon (8-5).

Total birds counted reached 958,439, of which 391,011 were classed as "oil vulnerable." The report said: "Our best estimate of the direct injury to marine birds...from...*Nestucca* is that 55,912 birds died. A total of 10,336 birds were recovered in Washington, 1,027 of which were successfully rehabilitated. It is estimated that 5,015 birds...beached in Washington were not recovered. We estimate that an additional 41,588 birds were carried northward with the slick or were killed as the slick drifted toward Vancouver Island. Of these birds, 10,000 are believed to have been beached,...of which 3,137 were recovered, and a total of 31,588 birds were lost at sea...."

Taking into account various uncertainties, the Ecological Consulting report calculated mortality using a randomization procedure and concluded "there is a 95% chance that the total mortality is in excess of 47,477...and a 5% chance that it is greater than 68,492."

Personal contact with Canada Wildlife produced a figure of 3,568 dead birds recovered on Vancouver Island, somewhat higher than the 3,137 figure in the Ecological Consulting Report, although perhaps 431 is not a significant difference. Only 33 live oiled birds were recovered in Canada, of which 19 survived until release. In Washington, 1,027 birds were released after cleaning.

What struck me as curious about the Ecological Consulting Report was that in figuring the bird population loss, the 1990 survey total was used as the basis, rather than adding the numbers of birds known to have been victims of the 1988 spill. The EC Report figured total mortality of 55,912 as 5.8 percent of the total bird population. Since murres were approximately 30,000 of the birds affected, the report noted that their mortality represented 11 percent of the murre population.

Add the birds killed by the oil spill to the total population of vulnerable birds, the mortality figure becomes 12.5 percent of "oil vulnerable" birds or 5.5 percent of "all seabirds." And if one takes the aerial survey's figure of 296,859 murres counted/estimated over the continental shelf and adds the 30,000 believed to have died after the spill to give a total of 326,859, the result is 9.2 percent mortality. That, of course, assumes a steady population, ignoring losses and gains from normal mortality, immigration and emigration and fledged replacements from breeding pairs, which lay but one egg annually.

My curiosity concerning the numbers was not supported when I inquired of my older brother, J. Dan Webster, Ph.D., retired college biology professor/ornithologist. "I think they used the 1990 January figure as a basis for these reasons, which seem valid to me," he wrote. "1--They are the best census figures (almost the only ones) available for the season. 2--They wanted to be very conservative on losses. 3--Natural mortality and increase figures on the populations studied, as well as the effects of man-caused incidents, would be wild guesses without far more information than is presently available. However, they should have discussed the use of their 1990 census to measure 1988-90 losses."

When I checked with U.S. Fish & Wildlife people in our area, a new factor was revealed. The number of breeding pairs of murres along the Washington coast has been falling precipitously, from 30,000 a few years ago to less than 800. And the murres represented about 30,000 of the birds that died as a result of the *Nestucca* spill.

Yet another factor was mentioned, the possibility that many of the victims of the 1988 Christmas oil disaster were not resident birds, but young murres bred the previous summer in California that moved north in the fall looking for "greener pastures" and on this occasion found oiled waters. Similar immigration might occur from Southeast Alaska or Canada.

Wouldn't it be convenient if life were simple? But that seldom happens. In this situation, the puzzle is keeping the naturalists busy searching for reasons why, and for solutions.

Playing games with figures can be interesting, but in this case only helps explain the initial statement, that such answers are not simple.

Final reports on known bird mortality and releases are more dependable. Wildlife Canada, in addition to its total of 3,568 dead birds along the west coast of Vancouver Island, provided further figures. Birds released after cleaning and rehabilitation included one common loon, three red-throated loons, two horned grebes, two white-winged scoters, one mew gull, six glaucous-winged gulls, seven common murres, one pigeon guillemot and three ancient murrelets.

Among a group of 856 dead birds identified by species, there were 356 common murres, 42 percent of the total. The rest included 30 other species, as follows: Cassin's auklet, 274 and 32 percent; scoters,

6 percent, 55 unspecified among white-winged, surf and black; gulls, 4 percent with 34 including herring, glaucous-winged, California and kittiwake; loons, 3 percent at 25 including common, Pacific and red-throated. Then there were 10 northern fulmars, 12 grebes (western, horned and red-necked), 13 cormorants of various species, 12 marbled murrelets, 21 ancient murrelets and 15 parakeet auklets, plus one to three birds of several additional species.

U.S. Fish & Wildlife Service's Ecological Services office in Olympia, headed at that time by Don Kane as field response coordinator, compiled a complete report on the live and dead birds collected from Washington beaches. Rather than reproduce the spreadsheet, here are its essentials.

The largest single group of live birds was the common murre, of which 706 were released after rehabilitation, but 6,067 had died. Second largest group was the scoters, with 281 birds surviving until release and 824 others dead. Of these, 231 white-winged scoters made it to freedom, plus 24 surf scoters, 7 black scoters and another 19 not specified as to species. Third largest group of birds affected was the grebes, of which only 31 survived out of a total of 748. Of those survivors, 27 were identified as western and one as a horned grebe, while three others went unidentified but probably were westerns.

I recall spending a shift tubing slurry down the gullets of several pens filled with clamoring western grebes, and was sad to learn that so few of them had survived. They're even more feisty and independent than the murres, with a ruby-eyed glare that's almost hypnotic. Their speedy beaks can stab the unwary, whether finned or footed.

As to the remainder, the list is alphabetical, first those dead, then a slash (/) followed by the number that survived to release. Cassin's auklet, 72/0; parakeet auklet, 1/0; rhinocerous auklet, 34/1; bufflehead, 8/1; canvasback, 0/1; coot, 1/0; cormorant, 16/0 (including one Brandt's cormorant); crow, 4/0; dunlin, 4/2; fulmar, 53/0; goldeneye, 3/1; Canada goose, 1/0; guillemot, 1/0; gull, 116/0; glaucous-winged gull, 2/0; herring gull, 8/0; mew gull, 2/0; kittiwake, 23/0; loon, 63/1 (the survivor was probably a common loon); Arctic loon, 13/0; common loon 2/0; Pacific loon, 1/0; red-throated loon, 4/0; mallard, 1/0; merganser, 1/0; murrelet, 11/0; ancient murrelet, 4/0; marbled murrelet, 2/0; oldsquaw, 2/0; petrel, 2/0; storm petrel, 9/0; pheasant, 1/0; puffin, 4/1; horned puffin, 2/1; tufted puffin, 4/0; sandpiper, 1/0; shearwater, 13/0; scaup, 1/0; tubenose, 5/0; unknown, 648/0, and parts, 363.

Of the total birds checked at Ocean Shores, Neah Bay and by rehabbers, 9,309 were dead or died in custody, and of the 1,027 released, 34 of them were at Neah Bay, 72 by rehabbers and the remainder either at Ocean Shores or at the Nisqually Wildlife Refuge 10 miles east of Olympia where the Nisqually River flows into Puget Sound. Of the 3,092 live birds collected from the beaches, the 1,027 released represent 33.2 percent; 66.8 percent died.

The long-term population trend of marine birds off the Washington coast since *Nestucca* is somewhat better known, largely because some of the funds received in the court settlement have been dedicated to the purpose.

And the birds? Well, they continue to suffer repeated damage to their population and environment, but some limited help is coming (see USF&WS *Nestucca* Restoration Plan, 1995).

Ulrich W. Wilson of the U.S. Fish & Wildlife Service, stationed at the Coastal Refuges Office in Sequim, Washington, has related the fluctuating populations of murres and two species of cormorant to El Nino events over the years. Wilson also mentioned the effects of oil spills, gillnets and even Navy bombing practice (since discontinued) off the coast (8-6).

In a telephone conversation, Ulrich reported only 565 common murres were counted during the 1993 breeding season between Cape Flattery and Copalis Rocks, plus another independent population of 205 at Tatoosh Island. He said he believes the 1992 El Nino was in part responsible, then added that another major factor in the drastic population drop is believed to be seining for baitfish. Wilson said fishing vessels harvest the same schools of small fish that are being caught by murres, with the result that one seine boat in a single season tallied 1,200 dead murres. There are some 10 boats catching baitfish in the area, Wilson said.

"The future conservation of seabirds requires that we better comprehend and manage our own impacts on these species," wrote Wilson in his Condor article. "This can only be accomplished with a prior knowledge of how environmental factors cause variation in seabird populations."

Eliminating the devastation of marine bird populations by repeated oil spills and seeking to reduce bird fatalities from bait fishing would be steps in the right direction. These are events subject to control. Perhaps the numbers of murres and other offshore birds can slowly be restored. But at present, there is a lot of ocean out there devoid of much of its former seabird life.

Dave LeBlanc, Tofino, B.C. , volunteer spill response leader.
(Photo by David Webster)

9

Canadian Outrage Gets Results

Their beaches fouled by oil without prior public warning, ordinary Canadians reacted first as volunteer cleanup workers and secondly as outraged citizens burning with indignation at slow and inadequate bureaucratic response.

"Step out of the pickup and the West Coast grabs you before your boots hit the sand. *Big* ocean. *Big* trees. That distinctive tang of spray and rain forest as the Pacific rollers come booming in.... (9-1).

"Up by the tide line, a grizzled man in a yellow slicker pokes among huge logs spattered with globs of congealed bunker-C oil. He's looking for dead birds, hoping to deny scavengers their poisoned flesh.... No one knows how much oil is still out there waiting to come ashore.

"'The next storm will tell,' says this Vancouver merchant seaman, who has seen a new tide cover this just-cleaned beach with reeking oil. 'What about the Coast Guard, can't they help?' the mariner is asked. His jaw clamps shut, his face hardens and he turns abruptly away. 'They wouldn't believe me,' he says, finally. 'They said it was nothing. Even when I put a bag of oil in their hands.'

"Persistently downplayed, virtually ignored from the start, this disaster on the west coast of the (Vancouver) Island has been declared officially over a few too many times. For the beleaguered residents of Tofino, already reeling from a four-month fight to keep two big multinationals from clearcutting Clayoquot Sound, the only part of this latest crisis that's 'over' is the government's credibility....

"This must be every Canadian's favorite place. After seeing TV...of oil...on Long Beach, more than 2,000 people responded. Some donated food and cheques. Others drove out from Ontario, Manitoba and northern Saskatchewan to help clean up the mess.

"Feelings run high among volunteers crossing the oil-stained carpet into the dining room of Tofino's Maquinna Hotel. Their food vouchers are good for a $14 dinner, courtesy of the Provincial Emergency Program. TinWis Lodge will put them up overnight. They'll get bag lunches and new assignments at the Friends of Clayoquot Sound volunteer centre tomorrow.

"The Maquinna's overworked waitress would agree that the off-season is over in this town at the end of the road, where a single winter's incessant downpour is enough to send most newcomers packing. But the harbour--usually raucous with over-wintering cormorants, grebes, ducks and loons--is deserted and silent.... 'I cry myself to sleep sometimes,' says...Catherine 'Cody' Gregory.... She closes her eyes and has visions of bald mountains lapped by sloshing oil. 'Yeah, people said it could come this way,' Cody recalls. But it was

Christmas. Then New Year's. 'We assumed the officials would monitor the situation. And in our secret hearts, we hoped it wouldn't happen.'

..."First affected, last consulted, the 14 bands of the Nuu-Chah-Nulth Tribal Council find traditional food sources wiped out overnight. An aircraft has to be chartered to fly provisions from Gold River to Friendly Cove. 'They didn't realize this is our home,' says a frustrated Richard Lucas.... It has taken the tribal council a long time to convince the Coast Guard to rely on the inhabitants who know this wild northern coastline best. Now things are coming together. Young tribal workers seated nearby type updates and take calls from isolated communities where 120 hired native Indians are out walking the beaches daily....

"The natives have evolved here with the living ocean, mountains and sealife for thousands of years, Chief Simon Lucas tells a hushed audience at (an oil spill) assessment meeting. 'Our grandfathers roamed these waters in 40 and 50-foot canoes.' Tracing a broken web of interdependent lives--from tiny filter-feeders to the mink and raccoons of the forests--the chief slams the government experts' numbers and evasions. The impassioned chief challenges official claims the spill is 'insignificant.'

"'That's a scientist's word! Intelligent people are destroying this world. This whole dominant attitude has to go.' But we allowed it, the chief concludes. 'I drove here in a car. We heat our homes and draw our water with oil. It will mean a change in lifestyle. Otherwise, I guarantee we will be doing this over and over again'."

These representative attitudes are but samples of Canadian feelings and responses to the oil that an unanticipated Southwest gale deposited on Vancouver Island beaches. Experts on the NOAA staff had expected only a light spattering of small oil globs to be deposited as the widely-spread spill drifted northwestward along the ocean coast. Once again, nature outsmarted the experts. Combined with the lack of preparedness and the unwillingness of the Canadian Coast Guard to request Army participation in cleanup, the public reaction was predictable. There were threats to haul the oily beach debris back to Washington State, calls for an international investigation to pin the blame for Canada's injury.

"We just could feel the outrage from the Canadians," recalled Ron Holcomb, public relations person for Washington State Ecology. "(They) seemed to feel that the decision to pull the barge (*Nestucca*) out to sea was purposely done to save Grays Harbor at the expense of Canada."

As the writer pointed out to Holcomb, that was inadvertent misunderstanding for want of information, both in British Columbia and Washington State. Canadians mistakenly thought that Lewey Kittle, on-scene coordinator for Ecology, had made the decision, when all he did was advise the Coast Guard of the facts regarding the ecological sensitivity of Grays Harbor. And even the Coast Guard, which advised the skipper of the tug *Ocean Service* to take the leaking barge offshore, did not make the decision. Skipper Charles May had already realized he could not tow the barge into the harbor against an outrunning tide with but minimal steering control, and was headed seaward as the only recourse after re-establishing his connection to the barge by employing the Orville hook.

In the end, political reason prevailed and international confrontation was avoided. Governor Booth Gardner of Washington State and Premier William Vander Zalm of British Columbia met to calm the troubled waters between longtime friendly neighbors. They named a task force of British Columbia and Washington State people to cope with the problems created by *Nestucca* oil crossing the international boundary.

HOLIDAY SURPRISE

There is but one paved road across Vancouver Island to reach Port Alberni at the head of Barkley Sound, Ucluelet at its entrance and-- some 25 miles north past Long Beach--Tofino, at the south entrance to Clayoquot Sound. Except for a coastal stretch from Victoria west to Port Renfrew at the south end of the 282-mile-long mountainous island, this is the only paved highway on the ocean front. Other scattered communities along the western coast are small, and few can be reached by cross-island roads. Most people on Vancouver Island live along its opposite shore, facing north and east toward the British Columbia mainland.

Vancouver Island's western shore is on the flyway for birds migrating along the Pacific Coast, and includes breeding areas for large colonies of seabirds. Whales are common, and Canada's only sea otters call it home, as do the Native people who have lived there for thousands of years.

The Nootka area 50 miles northwest of Tofino took an early position in West Coast history when Captain James Cook from England visited on the voyage that ended with his death at the hands of Hawaiians in 1779. Juan Perez had first visited in 1774. The local story on how Captain Cook came to call the place Nootka is that he waved his arm in trying to get the local name, and the Native people mistook his circling gesture to mean he wanted to go around the island. That is the meaning of the expression "Nootka," but the English naval officer took it to mean the location. Nootka later was used as a protected anchorage by many traders bartering furs from the Natives to sell in China.

The cross-island highway starts from Parksville some 20 miles north of Nanaimo, terminus of the provincial ferry route from West Vancouver. Highway 4 is a two-lane road carrying little traffic as it winds through the forested mountains of the island's backbone and drops down to Port Alberni, then courses the north shore of Sproat Lake and on over the watershed divide to Kennedy River, winding through and over hills, finally down to Kennedy Lake and on to a junction with the coastal road. The east end of Sproat Lake has the last residences for nearly 50 miles until the coast, but the scenery is spectacular.

The only negative aspect is observing how the forest's removal by clearcutting has caused major erosion and landslides off the steep hillsides, blocking side streams and sometimes washing out the highway. But that is helpful in preparing one for the unwelcome concept of this beautiful ocean shoreline polluted with spilled oil.

As the *Nestucca*'s discharge arrived on the beaches of Vancouver Island in British Columbia, a different reaction developed than in Washington State. There are in truth two stories of Canadian response to the *Nestucca* oil. One version is in official government reports, and the other was the experience of the volunteers, who were coordinated--perhaps reluctantly--by Dave LeBlanc of Tofino.

LeBlanc was proprietor of a shellfish farm on a saltwater inlet east of Tofino. He is a former employee of Coast Guard Canada at the Tofino Lifeboat Station, also a fisherman and beachcomber. An amateur radio operator, he keeps the receiver in his apartment tuned to the lighthouse channel to stay in touch with marine conditions. More recently, he has operated a water taxi out of Tofino.

When LeBlanc heard of the *Nestucca* spill, his knowledge of coastal and offshore currents caused concern that the oil might reach Canada. Such awareness was not general along the lightly-populated coast on the western shore of Vancouver Island, for a couple of reasons.

First, it was the Christmas season, when people have other concerns, and there was no public announcement of the imminent likelihood of oil on the beaches. Official belief was that it probably would not happen; and even if it did, the oil would merely "spatter" the beaches with droplets of oil. Secondly, by happenstance, the Canadian cable television channels serving the west side of the island were changing the source of their programming from Seattle broadcast channels to a satellite supplying programs from the East. Changeover occurred at the end of the calendar year, so that when oil actually hit Vancouver Island, the event was not reported over television news. Easterners could not be expected to care about an oil spill on the far west coast. The *Nestucca* spill, an ecological disaster locally, wasn't enough for more than momentary attention back east. Not until the *Valdez* oil spill in Alaska a couple of months later, involving some 40 times the quantity of oil from *Nestucca*, did national interest develop. The earlier spill, by then mostly cleaned up, was ignored.

But there was more to the January situation in British Columbia, because of the history of events related to logging. The popular view was that Tofino is largely populated by "tree huggers" and others with ecological concerns, while Ucluelet at the south end of Long Beach is made up of loggers and others dependent upon resource exploitation for their livelihood. As with any such generality, that view is neither altogether correct nor universally shared. Ucluelet's populace includes quite a number of native tribal people, and a considerable share of the economy is related to fishing and tourism, plus the Coast Guard, not just logging.

As the Canadian Coast Guard report on the incident puts it, in "Setting the stage,...the villages of Tofino and Ucluelet...at either end of Long Beach...form a dynamic mix of natives, environmentalists, loggers and fishermen. Tofino...(has) a population of 1,000 whose...main income is...from tourism and fishing. Ucluelet,...with... 1,500 (has) logging and government work...50 percent of the economy; the other 50 percent...from tourism and fishing. During summer, tourism is...vital,...worth $14.3 million annually.

"Residents of the area were understandably upset when they found...oil on the beaches," the Coast Guard report continues. "The *Nestucca* spill cleanup was a new experience for most people. This was the first time native groups and large numbers of volunteers have become involved in oil spill cleanup operations, ...and adjustments had to be made to incorporate them...."

LeBlanc and fellow volunteers recall things differently.

There has been for years a strong public reaction, complete with demonstrations, picketing and arrests, accompanied by court proceedings and loud public debate, against timber policies which have allowed almost unrestricted and destructive clearcut logging by large corporations. Tofino has been a center of this anti-logging agitation.

The stretch of coast from Ucluclet to Tofino includes 15 miles of Pacific Rim National Park lands. The park also includes the Broken Islands group in the mouth of Barkley Sound south of Ucluelet and another 40 miles of mostly roadless and uninhabited coastline southeast to Port Renfrew. That is still about 50 road miles from British Columbia's capitol city of Victoria, which is at the southeast tip of Vancouver Island across Juan de Fuca Strait from the Olympic Peninsula in Washington State.

Coast Guard Canada's station at Amphitrite Point, just south of Ucluelet, is part of a unique vessel traffic control operation with the United States Coast Guard's Thirteenth District. The system works similarly to air traffic control, but its directives are only advisory, not mandatory. All vessels longer than 20 meters (66 feet) approaching Juan de Fuca Strait are required to check in by radio with Amphitrite Point, providing information as to their identity and destination.

Inbound vessels are tracked by Coast Guard radar, communication and control being handed over by Coast Guard Canada to the U.S. Coast Guard three miles inside the entrance to the strait. For outbound vessels the reverse applies.

L.W. (Larry) Pokeda, Officer in Charge of Vessel Traffic Services at Amphitrite Point, explained that the system was created in 1978-79 as a joint management agreement between the two coast guard services. To his knowledge, nothing comparable exists elsewhere.

Pokeda said vessel traffic control is complicated by fishing vessels offshore, which can number 250 during summer along the west coast of Vancouver Island. Some of those fishermen are on foreign boats, and their command of English is limited so that accurate communication is difficult. Blips on the radar screen can't explain what a vessel may be doing, or when its course may change, Pokeda observed, so that guiding a large commercial vessel such as an international container ship or an oil-laden tanker causes concern.

Sinking of the *Tenyo Maru* July 22, 1991, provided an example of the difficulty. The Japanese fishing vessel was involved in a collision with the Chinese freighter *Tuo Hai* shortly before 9 a.m. on a foggy day. The *Tenyo Maru* sank immediately, and over the next month and more released an estimated 100,000 gallons of oil. Fortunately, most of it drifted south and remained offshore in calm weather. More than 4,000 birds, mostly murres, fell victim to that oil spill. But that was two and a half years into what was then the future, so let's return to 1988-89.

"On the evening of December 31st (1988) I heard the lighthouse keeper from Carmanah Light Station say that he was getting dime to quarter-size tarballs on the beach," recalled LeBlanc as we talked (with tape recorder) in his apartment.

Webster: "Where is...Carmanah?"

LeBlanc: "About 12-13 miles north of Cape Flattery (northwest-ernmost point of the south 48 United States) on the Vancouver Island coast, about 50 miles southeast from here. Naturally I was quite concerned, because if it had come all the distance from Grays Harbor to Carmanah, the potential of continuing north along the coastline was definitely there."

Webster: "Were you aware, Dave, of the Davidson Current setting northward in the winter?" (9-2).

LeBlanc: "Oh, absolutely. I've spent a lot of years off the coast, both beachcombing and commercial fishing, and I was...waiting for a response from the Coast Guard. Intercepting their conversations, I knew the American Coast Guard didn't think there was going to be any significant oiling in Canada. The general drift was they didn't think it was much of an issue.

"Then in the following days, the lighthouses would report oiling further along the coast. Pachena Point Lighthouse reported tarballs coming ashore...around the first of January. I'm staying home and listening, and the people involved were still in Victoria or Vancouver. There was no mobilization. I hadn't really discussed the matter with anybody. Because the Seattle stations were no longer being carried over our satellite, we weren't getting any information as to the oiling along the coast. Nobody here had any idea that there was even an oil spill occurring. I knew,...but the residents here didn't.

"So when oil came ashore here on January 3, people were absolutely flabbergasted, because they had heard nothing of it. If you think about an accident that occurred on December 22, and oil is now washing ashore here, you're looking at 10 days later. That...is...distressing. Everybody knew, the agencies were all notified, but they chose not to react. Probably because it was Christmas-time...had a lot to do with it. People were more concerned with whether they were going to have cranberries--discussing whether it would be red wine or white wine."

Webster: "A lot of things going on beside concern over an oil spill."

LeBlanc: "Exactly. Especially when the information they were given was...the likelihood is that you're not going to get it. So nobody thought anything of it. Except for myself. I was quite concerned.

"The first thing that came in was birds. Before we got any major oiling with patties and continuous coverage, we were getting the birds. On the very first day,...January 3, a lot of the residents and parks staffers mustered up whatever they had to get on the beaches and get to work. Before...any orchestrated organization, people were just going out on their own on the beaches."

Webster: "How did they get the word, if it wasn't on TV, it wasn't on general radio news?"

LeBlanc: "Somebody came walking into the (Common Loaf) bakery and says, 'Hey, do you know there's a bunch of oiled dead birds on the beach. Oil all over the place. Where's the government, where's the local....?'"

Webster: "You got it from the grapevine,...word of mouth."

LeBlanc: "That's right. Nobody was warned of anything. I was one of the only people in Tofino that had an early warning. So people were just going out and doing things.

"Now...I went down and had a chat with Brian Stevens, who was my officer in charge at the Tofino Lifeboat Station, and asked him what kind of information they had. The only thing he knew was that Larry Slaght was the on-scene commander, and they were out having a look around." LeBlanc, who was a part-time employee at the lifeboat station, had been laid off in December.

"It was rather unfortunate...for the lifeboat station, because their sole purpose is lifesaving, and they were starting to be inundated with calls. 'Hey, there's oil on this beach, there's oil on that beach. What are you gonna do, where are you going?' They weren't advised of anything, and they were facing the brunt of local pressure. It was kind of unfair.

"Anytime we'd call any of those (government telephone) numbers you'd get someone who'd say, 'Oh, well, is there oil on there?' And that's the way the whole incident went. Anything that we discovered was always news to the officials.

"Meanwhile, these lighthouse stations are reporting continuing oiling, day after day. Park staff in Pacific Rim National Park didn't have any equipment to speak of, so over the next couple of days every garbage sack in town got bought up, and...a lot of carpets were getting 's'oiled.'

"I had to get out and see this, to an area where nobody had been, and...I chose...Ahous Bay on Vargas Island. I'm very familiar,...I dig razor clams, beachcomb there. It's a beautiful remote beach, large, nobody goes there. It's maybe 8 or 9 miles by water, but I flew in that day. Landed on the beach with two other individuals, not counting the pilot, and arranged for the plane to go back to Tofino and pick up three others, because I knew we had a lot of work to do.

"That day, January 4, we collected over 200...dead birds, and we didn't have time to pick up the oil itself. There just wasn't enough time in the day to do all the work. We...got as many birds as we could off the beach, so predators wouldn't get at them."

Webster: "Did you record what kinds of birds these were?"

LeBlanc: "Common murres, Cassin's auklets, scoters and loons. A lot of them had been eviscerated by predators, mainly eagles and seagulls. We were seeing a large number of marginally oiled seagulls from picking on them (dead birds). Impossible to catch. We did what we could, for as much daylight as we had that day.

"So we went back to Tofino, and I met with Lindsay Armstrong and we had a discussion. 'What are we going to do about this, Lindsay?' And he said, 'Well, who are we going to call?' We tried to call the provincial emergency coordinator, and he was not available. I think he was trying to get some bureaucratic wheels turning. We got no response from our local emergency response coordinator. Nobody in the Coast Guard that we could talk to, of the people in charge.

"On January 4 the Coast Guard advised Brian Stevens to go down on the beach and have a look. They'd already received scveral reports...of oil on the beach. They chose not to believe them."

Webster: "How much oil was actually being reported?"

LeBlanc: "Oh, large patties, covering the entire length of the beach. They tend to call it minor oiling because, you know, it wasn't covering the entire beach from end to end a foot deep. When I see oil on Chesterman's Beach (just south of Tofino), I don't consider it minor no matter what. It's not minor to clams and everything else on the beach. And there's millions of migrating birds that go by there, an extremely important area for them.

"And still no sign of an elevated response. There was only a little bit of looking around. The day we were on Vargas we saw a helicopter pass overhead. They didn't...even appear to look, just went full airspeed overhead. I guess they went somewhere...further north to take a sample. They obviously saw us on the beach picking up oiled birds. ...Then we saw them going back the other direction, same speed, same altitude. I understood (later) it was Martyn Green, the consultant hired by Sause Brothers Ocean Towing to represent them in Canada. The polluter at that time was...saying, 'Hey, how do you even know it's our oil. We're not going to do anything until (we know) it's ours'."

The Canadian Coast Guard report, however, noted that on January 3 the Washington State Department of Ecology laboratory confirmed a Carmanah Point sample as being from *Nestucca*, and January 5 the previous day's samples were confirmed.

LeBlanc: "To me that seems reprehensible, because we've already got 50 miles of Canadian coastline...contaminated by oil, and they've still trying to decide whose it is.

"So what happened in that meeting with Lindsay Armstrong turned out to be very important.... We knew oil was going to continue to come ashore, and we knew a lot of people that live in remote locations (further northwest up the island). Now if there was no response in obviously very public areas, we knew there was probably going to be no response in these remote areas.... People on Vargas Island, Flores Island, Estevan, Hot Springs, Nootka, all along the coastal area...were going to wind up doing the work themselves.

"What we decided...was we wanted them to report to us the... work they...do, so in the event of any legal repercussions we would have...centralized, credible...evidence to bring to court. That was our purpose.

"Lindsay said to me, 'Well, you might ruffle a few Coast Guard feathers, because you are a Coast Guard employee.' And I...had to decide whether to be an environmentalist or an underling of the Canadian Coast Guard. Environmental issues are of great importance (to me), and...I took a $50,000-a-year job and threw it out...in order to help good old Mother Earth. It's kind of been a difficult process, but...my conscience is...better for it.

"And we decided, okay, this is what we're gonna do. We got some charts, we got a diary, we got a phone and some extra phone wire, and we asked permission to set up a coordination center in the bakery (the Common Loaf Bakery across from the Tofino Post Office). It's kind of the hub of the environmental movement.

"January 5th, the next day, turned out to be an absolute flurry of activity. A lot of requests were coming in, plus we were sending people out to get reports."

Webster: "These people you were sending out...were local volunteers?"

LeBlanc: "These were volunteers who were coming, saying 'What's happening, where is everybody, what's going on?' So we said, 'Instead of worrying about bashing heads right now, you go down to Tonquin Beach, and you go down to MacKenzie Beach, you go to Chesterman's Beach, see what you see, report...back...what you find.'

"By this time we had decided, okay, they're not going to do it, we're going to do it."

Webster: "On January 6, the next day, the Coast Guard Canada report says 'the Sause Brothers contractor hired eight labourers in Tofino and they began cleaning up oil in Pacific Rim National Park'."

LeBlanc: "Well, that's not true. A...Tom Johnstone,...the first agent from Sprayaway, subcontractor...under Burrard Clean, the contractor hired by Sause...came to the coordination center, and that's where he spent the day. He didn't have anywhere else to go. He came, he sat with us,...and he got some kind of idea what was going on.

"I didn't like the man from the first moment,...because first off, he wasn't going to give us the radio frequency. We had a standard setup,...listening to the park, the Coast Guard, helicopters, the police, hearing all the reports, getting all the information. I said to him 'What is your frequency?' and he said 'No, you can't have it, that's a corporate secret.' And I said 'Fine, you don't have to give it to me.' Then I asked...'What's your experience cleaning up major oil spills?' He said 'I've never done it before.' I asked, 'Well, where's your tools, where's your equipment?' And he said, 'We don't have any. We have a truck coming with some stuff, but we don't have anything.'

"That was really distressing. Seventeen days after the incident, and a contractor doesn't even have tools? Johnstone said he was going to hire eight guys. The bakery is full of people...looking for work. First he proposed cutting up little pieces of paper to draw for the eight people.

"And I said, 'Come on, Tom, that's stupid. You want eight people, just tell me and I'll pick eight of the best people for you.' He thought about it a minute, and said 'Okay.' So I said 'You, you and you, you got a job. You go with this guy.'

"Meanwhile I got 34 people out working the beach already. He hired eight. Oh yes, he wanted one guy to have a first aid certificate for his crew, and I believe one person did have some survival-oriented first aid. And he wanted them to have their own raingear,...shovels and rakes.

"I thought, okay, I'm cooperating with him. It's slow, but at least we're moving. Then Johnstone said, 'Okay, well, I'm going to go...and have a meeting, so you guys just wait here.' They sat for the remainder of that day. They'd been hired, but never did...work.

"It was also that day, January 6, that I met Martyn Green, the consultant hired by Sause Brothers. He spoke first of our 'sterling job,' and said he thought it would be good if we were '...in the process (with them).' He also recommended that I take on the role of volunteer coordinator. Some people might say that it's a self-proclaimed position, but it really isn't.

"Martyn Green said 'Okay, you seem to be doing this, you be the volunteer coordinator, and you'll be the guy and everybody will do it, and so you're not self-appointed.'

The Coast Guard report for January 6 includes additional items: "Media and public concern mounting over the oil spill. Threats to sensitive resource areas such as shellfish...and sea otter(s)...are identified for monitoring.... Department of Fisheries and Oceans closes two shellfish areas as a precaution to avoid contaminated product reaching the market. Canadian Coast Guard helicopter... reports oil slicks up to eight km. long, 50 metres wide, eight km. offshore parallel to shore and extending from eight km. north...to 30 km. south of Tofino.

"Colin Hendry, CCG's Regional Manager of Marine Emergencies, replaces Larry Slaght, CCG Victoria District Manager and pre-designated On-Scene Commander for Vancouver Island, as the On-Scene Commander."

The following day, January 7, by Coast Guard request, the contractor increased his work force to 47 people, and the report noted more than 100 volunteers were involved in cleanup.

CCG command centre group and OSC Hendry arrived that day at Amphitrite Point, Ucluelet, establishing operational headquarters in a trailer on the Coast Guard grounds. Three telephone lines were installed, plus connections to the CCG radio station. Various other groups, including the Nuu-Chah-Nulth Tribal Council, Environment Canada, Department of Fisheries and Oceans also operated there, at first in the garage and basement, but later utilizing the former lightkeeper's house with a total of nine phone lines and a fax machine.

LeBlanc commented on two other entries from the Coast Guard log for January 7, the first a notation that two CCG sea trucks started cleaning in the Broken Group Islands at the mouth of Barkley Sound.

"It was a postage stamp response. It makes some sense, and I don't disagree, other than that they didn't really...pick up any oil with those trucks, because, A, they didn't have enough equipment; B, they didn't have enough people, and C, the boats are impractical for the purpose. So they were starting to throw a lot of money at the problem, and it's not helping. As it turned out in the Broken Island group, it was the Natives (tribal folk) who cleaned up the islands."

Webster: "It sounds like the Native people did an awful lot and got very little credit."

LeBlanc: "Yes, what we didn't do, the Natives did. Actually, they did get a lot of credit. At that time...the Coast Guard was finally starting to realize that 'Hey, maybe we've got a bigger can of worms on our hands than we...thought.'

Webster: "You were on-scene commander locally?"

LeBlanc: "I was on-scene commander through the entire incident. I was the guy that everybody came to see. I was...reporting to fisheries, and as well as cleanup, we were also starting to get into surveillance. I was sending people into areas.... I wasn't satisfied with the Coast Guard's attitude not to do anything unless somebody told them there was oil,...and that was really aggravating.... They were only waiting for oil to come onto the beach.

"The thing that will happen is, the oil will get taken back off the beach...by the tide, only to be dragged through the marine environment again to be put up on the beach again,...multiple contaminations with the same oil.

"I think the polluter and the Coast Guard were both...negligent, because containable and recoverable oil...wasn't. It was allowed to re-pollute.... It's...called re-oiling.

"(In response to) Western Canada Wilderness Committee (intentions) to start...environmental...picketing if there was no more (government) action, the Coast Guard hired the lighthouse tender *Sir James Douglas*.... All of a sudden...aids to navigation were just forgotten. What was happening was...(that) equipment (the Coast Guard) had was not adequate, so they started putting other agencies into it."

Webster: "And their equipment then became available to help in the cleanup?"

LeBlanc: "Yeah. The lighthouse keepers were complaining...they were not getting any attention. They were doing their work, and helicopters were coming in and out, and (the crews were) eating their food and drinking their water and using their facilities, and pouring fuel--and the lighthouse keepers, on the front line, trying to accommodate all these people, (were unable) to get out for food and doctor appointments. It was...a nightmare...and they weren't very happy about it.

"The polluter then was told to start hiring more people... but his work force (was) 47 while we were dealing with more than 100 volunteers per day. On any given day we were outmanning them at least two to one, most often three to one.

"And we were starting to get reports that the (hired) work force was ineffectual. They would send all of them down to one beach, and either it had been cleaned by us previously...or the foremen...were unfamiliar with tides and oceanic conditions, they were making people clean up oil at high tide.... You know, it's really difficult to do that, when people are slipping on logs and everything. We (got) reports that the hired workers were just sitting around fires to keep warm and not really doing any work, or shoving stuff into bags to give the impression they were doing something to earn their 10 bucks (dollars) an hour.

"We were cleaning from 15 to 20 areas per day, and the contracting crew was cleaning one or two, maybe three beaches a day.

"Another thing happened on January 6. We had to move our operation into the volunteer firehall, because there was too much going on, and we were interrupting the bakery. The firehall has a lot more room, where we could lay out our charts on big tables, and space for our communications setup."

Webster: "Now when you say the contract crews were cleaning at high tide, do you mean your crews were cleaning at low tide or half tide?"

LeBlanc: "I didn't bother sending anybody out at high tide. There's no point. We were doing it in two shifts. Sometimes we'd have a morning cleanup, then there'd be a high tide in the middle of the day, and we'd get in to do a little more cleanup before the end of the day."

Webster: "How many hours of the tide were you able to work effectively?"

LeBlanc: "Oh, if the tides were right, you could get 10 hours. Once the tide started to get about half way out, you could get three hours, then when it started to come in you'd get another three hours, so on a good day you might get up to 10 hours; but there wasn't enough daylight, so we would be lucky to get eight hours in.

"By this time we were getting a lot of helicopters, with...experts wanting to go look at it (the oil on the beaches). But helicopters cost $650 to $800 an hour, and they're just throwing money away. They should have been taking cleanup crews in there instead of scientists. You only need scientists to gather legal evidence. There was still no real, organized cleanup effort. We still had no guarantees from the polluter or anybody that cleanup work was going to be done. But we were up to almost 200 volunteer workers per day."

Webster: "Do you have any idea how many Native people might have been working on your volunteer crews or under directions from this headquarters on cleanup in their areas?"

LeBlanc: "I think all in all there were well in excess of 100 Natives involved in the cleanup from the various bands. We'd phone up the coast and ask 'Have you seen any oil?' and they'd say 'No,' but even that is a report and I'd consider them helping as volunteers.

"We had sent warnings that oil was moving northward, and was potentially going to affect the sea otter colony at Bajo Point on Nootka Island. We sent a plane up to have a look, with Dr. Jim

Darling (whale expert) as an observer, and he reported a large slick off Bajo Reef. So we relayed that to the Coast Guard and Fisheries, which was news to them. But we were in contact with Nootka all the time.

"Now I'd been attending meetings down at Ucluelet every night, after getting up at 6 o'clock every morning to get to the firehall and deal with the front line coping with oiled beaches all day, then have to go deal with the 'generals' in the evening. Met with individuals from our provincial emergency program finally on January 7.

"I want to (mention)...the Provincial Emergency Program, and a fellow named Pat Harkness,...the guy in charge. These guys from PEP sitting there, one, two, three, in a meeting, said 'Well, we're going to help you guys out. You need some box lunches.' They appointed me to go find a restaurant, strike a deal and set it all up, order the lunches each day, distribute them and cross reference the records covering the volunteer workers. Meanwhile there's three bureaucrats doing nothing. It did take all three of them to deliver the check to pay for the lunches. Oh, yes, they also arranged to have dirty clothes washed for the volunteers.

"The PEP also decided to set up their own volunteer coordination center...at the golf course...halfway between Ucluelet and Tofino, and they recommended that everyone register with them. I...told them I wasn't going to send people out to register before they go to the beach. So we went on registering volunteers, and I think they registered maybe 35 altogether. We registered close to 1,000.

"When Harkness informed me that they had the coordination center set up on the golf course, they didn't have any bags, and they didn't have any gloves, didn't have any rakes, didn't have any shovels. So I gave them gloves and bags and shovels and all that sort of stuff, but I made 'em give me a receipt for it. I was getting really frustrated, and I wound up doing all the work for them. I said, 'You guys are bureaucrats, why don't you do it? My job is cleanup. I'm sick and tired of having to do all this stuff for you.' By this time he had five guys, and they started cruising around talking to everybody, but they were not really doing anything.

"The polluter still gave us nothing. Throughout the entire incident all we ever got from the polluter was 800 cartons of plastic bags. I'm serious. By that time we had already given out and filled several thousand bags. We were buying them at our own expense from the local stores. You couldn't find another garbage bag on the west coast of Vancouver Island, because we had cleaned out the supply.

"One reason I was...(trying to cooperate) was that people were starting to get fairly critical.... We've...an environmental group here, and they're quite radical in their views. I'm an environmentalist by every stretch of the word, but I'm not a radical, and I didn't want it to turn into a head-bashing ceremony. I wanted to get the place cleaned up. Some might not consider me a moderate, but I don't make a point of doing illegal activity, I'm not into civil disobedience or that kind of stuff. But it's...time to start lockin' horns.

"By this time the media was starting to pick up on this. They go to one beach, it's volunteers; they go to the next beach, it's volunteers. They come into town and everybody's going, "Where is the polluter, where is the government?' They had just thrown this in our lap.

"I was attending two meetings each night, first the REET, the Regional Environmental Emergency Team, then the Coast Guard meeting. The idea was to report on conditions (at the REET meeting), and make recommendations on cleanup to give the Coast Guard, which would take (them) to the polluter to work on. This became frustrating, because it took forever and a day for those guys to get going. The first couple of times I sat in a meeting, it was 'Oh, well, we're not going here yet, we haven't got our charts up.' And this was like January 8 or 9.

"They were set up in a garage. Cold. It was...deplorable conditions to have to work under. You go into a meeting and hear the on-scene commander say 'Oh, everything's under control,' and...it's really frustrating."

Webster: "Did most of the volunteers...come from the immediate area of Tofino/Ucluelet, or...from farther?"

LeBlanc: "Largely from the local area, but people came from all around. We had people from as far as Saskatchewan.

"I saw on the videotape that the Ministers (Provincial government of British Columbia) decided to show up. They...went down to the beaches and talked to volunteers who were working with garbage bags on for raincoats. They didn't have enough equipment. But they (the government officials) were...ignoring everything the volunteers were saying. They went to Long Beach, and you got six volunteers cleaning up the whole damn beach.

"...And they went on to a press conference, and said they realized the magnitude of this spill is a lot bigger than...they originally expected, so they were going to step up the operation. So...they hired the *George R. Kirk*, which is an icebreaker, at $65,000 a day, and it never removed one bag of garbage."

Webster: "What did it do?"

LeBlanc: "Well, it delivered some fuel in the Estevan area, because the helicopters were flying back and forth a lot, and...then they went...up north and sat and waited for oil to come. It never came. We finally got some surveillance going. But why didn't they just tell somebody in the Queen Charlottes (island group north of Vancouver Island) to walk down to the beach and report?"

Webster: "What would an icebreaker have been doing had it not been on oil patrol?"

LeBlanc: "Probably aids to navigation, servicing lighthouses.

"There began to be a lot of internal squabbling. Environment Canada was...having a battle with the Coast Guard...over...what they were supposed to be doing. They...had no solid terms of reference. This was January 9. Let alone getting out and doing something, they were still out looking around. It wasn't until January 11 that...everybody knew where they were...as far as cleanup was concerned."

For balance, let's switch back now to the Canadian Coast Guard's *Nestucca Oil Spill Report*. In its recommendations section, the report states: "An overall assessment of the *Nestucca* operation would be that it accomplished what was intended. Media reports were critical, but in fact once the response organization was in place the cleanup proceeded satisfactorily.

"...Another important positive factor...was the contribution of the native people of Vancouver Island. While the spill incurred their wrath initially, they joined in and were a major force in the cleanup. Without their assistance,...(it) would have been far more difficult.

"On January 1 oil began washing ashore at Carmanah Point and...it was January 9 before a full-scale operation was underway. Several factors affect the...time it took.... First was the erroneous information...from United States officials. It was ten days before the amount of oil was known (to Canadian authorities) and almost four weeks before the movement of the leaking barge out to sea was revealed.

"Coast Guard policy (is)...that the polluter is responsible for cleanup and should provide the initial response. The practice is to sample the oil to determine or verify the source before taking action or confronting the polluter. ...This...delayed response. January 1... samples were taken...(and)...two days later...confirmed to be from the...*Nestucca*. On the same day, January 3, oil hit Long Beach. Samples...were confirmed...January 5. The polluter hired a Canadian cleanup contractor January 5.... It wasn't until January 10 that the contractor had sufficient personnel (115 people) and equipment to deal effectively with the pollution.

"The Coast Guard was not as quick to respond as it could have been. Delays in assessment, verification, and the decision to direct the polluter to undertake the cleanup were factors in the slow response."

These highlights are accompanied by recommendations and further detailed discussion. The portion dealing with volunteers contains interesting variations in attitude from LeBlanc's.

"This was the first time large numbers of volunteers became involved in...oil spill cleanup...in British Columbia. Well over 1,000 individuals volunteered...with as many as 150 active daily. There were no policy guidelines or contingency plans...on...use of volunteer forces. Neither...CCG nor the Provincial Government provided comprehensive direction of this effort. Another major issue concerning volunteers was...safety. While...generally well motivated and industrious, lack of training, communications, and direction contributed to inefficiency in their efforts.... Unsafe work practices resulted in the need for CCG to rescue people stranded on beaches as the tide advanced. One...problem...was inadequate supervision of the volunteers. ...The CCG had insufficient staff to provide supervision for the large volunteer forces present.... Volunteers were generally critical of the effectiveness of the Government response and the...support provided to them.... Greater attention was needed in communicating with and involving the volunteers in the process.

"A reasonable argument might be made for excluding volunteers...from response operations (but) it is likely for any spill...in B.C. waters that large numbers of people will want to volunteer. (They) can be effective,...(but this) will require considerable planning."

Coast Guard recommendations include integrating volunteer activity in response exercises, but with CG authority to close areas to the public, identify appropriate tasks for volunteers, have the Provincial Emergency Program supervise, equip and get them to work sites, using the Canadian Marine Rescue Auxiliary.

Commenting further on involvement of the Nuu-Chah-Nulth (whose name means "all along the mountains") and other Native people in the cleanup, the Coast Guard report admits that "No one realized the potential of the communities to assist with...the operation. As the cleanup progressed,...Native volunteers... became part of an extensive cleanup group under contract with the Coast Guard. Their intimate knowledge of the environment and their deep concern for the impact of the oil made them ideal workers. ...The monitoring program...to ensure that the shoreline remained free of oil was also carried out by Native people."

The Coast Guard comments on lack of communication and supervision of volunteers caused LeBlanc to respond.

"We were getting problems with re-oiling...I would use my beachcombing experience, looking at the obvious, and I'd just say, 'We haven't had reports from here and here and here yet, but I know we're going to get oil...because of the weather or certain conditions like today (it was raining hard and blowing while we talked in LeBlanc's cozy apartment). And we'd send them out to the beaches with directions to wait for the oil and when it comes in, clean it up.

"I'm of the opinion that if you see oil, you pick it up. We had a considerable problem with re-oiling. You'd report oil on a beach, then the tide would come in and wash it all away again, and the helicopter would go look at it the next morning, and...come back in and say 'We didn't find no oil on there, what're you talking about?' And I'd say 'Well, there was oil there yesterday. It's back polluting something else now.'

"A lot of the cleanup was...sort of staged for us. I helicoptered several times. One day they took a rescue helicopter from Comox, a search and rescue Labrador (type) with two rotors. Stripped her of all rescue gear and took her to a beach that we'd started cleaning a week before. It was unnecessary, a complete waste of money. It was all staging,.... not a practical operation. A rescue helicopter is over $2,000 an hour. I could have taken that $2,000 an hour and made a lot better use of it, and done a lot better job of cleanup.

"To give you an example of why I was not satisfied with the (government) cleanup, we went to...Stubbs Island,...a mile north of Tofino. We had taken a thousand bags of debris off there. We had all these people--Sause Brothers representative, REET, Environment, Richard Lucas from the Native Council. We parked the helicopter on the beach, and I was the only guy brought a shovel. Randy Simmons from Environment borrowed it to dig strata, looking at deposition and erosion. In wintertime the beaches build up and in summertime they erode. Simmons dug a little ditch and said 'The beach is in sedimentation now.'

"Now if you think of the time and season, there was going to be only a few more inches brought in, and by summer...the erosion would be more than what would be added. So any buried oil was going to be exposed during summer. Well, I took my shovel back and leaned on it while those experts walked from one end of the beach to the other, and they came back all smiles. They said, 'We did a beautiful job of cleaning up this beach'.

"And then I started to dig right underneath their feet, and I isolated a big oil blob the size of a pickup truck. 'What the Hell do you call that?' They hemmed and hawed a bit, and I got into a discussion with the Environment official about whether it would biodegrade. 'Where is your baseline data to prove it's going to do that (degrade) here? You don't have it, do you?' And he said 'No.' And I said 'You have no experience out on this coast, and I'm telling you that this is going to be exposed.'

"He asked, 'How do you suggest we do that for all the coast?' And I said you've got to go to every beach and clean it out of there. He said 'That involves a tremendous expense,' and I said, 'Hey, that's the polluter's responsibility.' Meanwhile the Sause Brothers representatives are standing there going 'Oh, oh.' It was obvious they were concerned, because I had put them on the spot. They had all declared that beach clean, when it wasn't clean."

Webster: "You're saying in effect that if they looked at a beach and didn't see any oil, they declared it clean?"

LeBlanc: "And most of the time they wouldn't even land. They'd just fly over...and say 'That beach is clean.' You know, oil changes its characteristics after it becomes weathered. Even walking on the beach, you can't see buried oil if you don't know what you're looking for. If a guy like me can tell, surely an expert should be able to. But if a so-called expert can walk that beach with me, and I can tell there is buried oil and he can't, it doesn't give me a very high opinion of their expertise.

"I was out one day with Martyn Green, and he saw one rock covered with oil, and said 'Now I wouldn't clean that,' and I said 'Well, I'm glad you told me, because I'm going to send somebody in to clean that, because harlequin ducks winter here, and their distribution range is not wide. A harlequin stands on that rock, he's going to get oiled, and I'm going to have someone clean it up'."

Webster: "How would you go about cleaning that rock?"

LeBlanc: "We'd scrape it with a stick, a trowel, paper, get as much oil off as we can and put it in a bag. Even if we only get half a gallon of debris off a beach, in my book that's half a gallon that's not in the environment."

Webster: "What were the extremes of weather you encountered during the course of this...work over several weeks?"

LeBlanc: "Largely southeast gales, sometimes north."

Webster: "So good weather would have been...like today, rain with maybe 10-15-mile wind?"

LeBlanc: "Weather like today would have been pretty average. Bad weather would be storm-force gales, 60-knot wind, southeast.

"Getting back to REET, its role had been changed. They came up with a process called 'Surveillance, evaluation, cleanup and assessment.' They wouldn't react to anything if it was just a report. They'd say 'Okay, we've got to confirm that.' So on the second day they go and look at it. So, it's been surveyed, they decide it should be cleaned up, so they go to the evaluation team and the third day they'd say 'Oh, this is going to take six guys and 500 bags.' And...next day they'd organize the cleanup team to go there on day four. And quite often...either the oil was taken off the beach by the tide and carried somewhere else, or you'd have bad weather, and break that chain. And they would not pick up the chain. Some places never got cleaned. So we just sent volunteer crews in anyway the next morning.'

"We had a meeting of REET one night, and everybody agreed, the scientific community agreed that we just did not have enough resources to do this job. And this was after all the hype of stepping up the operation. I'd go in with maybe 10 priorities, and be lucky if we could get two on a given day. It was unanimous among the group that we just didn't have the resources. People from Environment Canada were saying 'this is the worst oil spill in Canadian history from the standpoint of wildlife mortality.' They decided to take it to the Coast Guard, but unfortunately it never went any further. We were just about on hands and knees, begging for more resources to do the job. The on-scene commander just sat there. I'm sure that inside he knew that he wasn't being given what he needed to do the job.

"Both Richard Lucas and Larry Baird, Native representatives, also spoke before the incident commander. Lucas said 'I'm sick and tired of coming in at 6 o'clock in the morning and waiting around all day to get into a helicopter at 2 o'clock to fly for half an hour. We're just wasting our time.' And everybody else on the emergency response team had the same idea.

"Finally I spoke up and asked, 'Why are you sitting on your hands?' And Colin (Hendry, the on-scene commander) told me to shut up. That was just about enough, because when I would be giving my reports, he would often walk out of the room, and...do other things, look like he wasn't paying attention. I guess they weren't treating me any different than anybody else. And I told them 'I'm tired of beating my head against you bureaucrats.' I was exhausted, physically drained. I had taken about as much as a man can."

Webster: "Did your volunteer work go on?"

LeBlanc: "Oh, yes, absolutely. We were...finding the funding and sending planeloads up to Nootka Island and other places. I had no intention of discontinuing my coordination. I just refused to go to the meetings. I was tired of working 16-17-hour days. This was January 19, and I'm finding it necessary still to step up the operation, because they're just not doing the job. I was no longer going to attend meetings at night, I was going to join the volunteers in the soup kitchen. So on January 20, as soon as we shut down the volunteer center (for the night), I went and had a drink, and it was kinda nice to just take time off. And I went to the soup kitchen, and I sat down and had macaroni and cheese.

"I was looking at the atmosphere of what was going on around here, and it was wonderful. Everybody...was doing something. I felt great being amongst all these people, because things were really happening. These people were cleaning up the beaches, they're cooking food, they came here of their own volition. It was just so organized, it was really nice. Such a refreshing change to be with people actually doing something.

"We were getting reports that the volunteer crew...(found) a contractor crew had been there, but acted like they didn't care. They'd fill bags and leave them on the beach, and the tide would take them away. We had a cabin where the volunteers were going to stay, and they had to hike a fair distance to get there, then they had to clean up the mess the contractor left that day. They had to re-bag, retie, move everything above high water. After the contractor left an area, the volunteers were finding class 5 contamination, continuous coverage. They collected hundreds of bags while they were up there at our expense."

"We were starting to get calls from the north end of the island by this time, way beyond Nootka. The Coast Guard had taken their effort from Barkley Sound and the Broken Island group...and shifted north. They completely bypassed this area. There was no Coast Guard cleanup in this area.

"The contractor was ordered by the Coast Guard to hire additional staff...January 18. He agreed, but two days later reneged, and cut back his involvement in cleanup activities. By that time a lot of the contractor crews were working at various locations up the coast, and we weren't seeing them in our area."

Webster: "He was doing strictly cleanup?"

LeBlanc: "Yeah, and that was even questionable."

Webster: "Maybe I should ask what other areas were you... involved in?"

LeBlanc: "Primarily cleanup. Putting oil into bags, marking them with tape for visibility and putting them above high water mark."

Webster: "Was that just oil and any sand that adhered to it?"

LeBlanc: "Anything that was...contaminated. Seaweed, oiled birds, whatever. Even oily styrofoam cups. By this time we were given sample bags and tags by the Department of Fisheries and Oceans, because we were picking up oiled crabs and things. We were basically doing specimens for them. We were doing bird mortality and species counts for the Canadian Wildlife Service. If it was oiled, we would collect it.

"In addition we were doing surveillance, organizing accommodations for volunteers coming from outside (the area), we were organizing rescue and rehabilitation of birds. We had a local restaurant that was closed for the winter converted into a soup kitchen, and any volunteer who worked that day was welcome...to eat, free of charge. We were starting to get food donations. People were watching TV and saying 'these poor volunteers, the government's busy locking horns with each other, the polluter's doing nothing for these guys, and they're doing all the work. Let's help 'em out.' There was truckloads of stuff coming up (the highway) for us. We were getting gloves, food from supermarkets, donations of garbage bags and a lot of things we needed, like shovels. The Ministry of Forests donated copious quantities of shovels for us. People were beginning to see the situation for what it was. There was a low level of response, and the volunteers were just carrying the load of this cleanup."

Webster: "You mentioned housing, and I might as well ask now, how were you handling that, was it in private homes?"

LeBlanc: "We had...different things going. We had people who had spare bedrooms, floor space, anything people could sleep on. At the local hotel we were getting meal allowances, special discounts to oil spill volunteers. Also we had an opportunity with the TinWis guesthouse, owned and operated by the Native tribal people. Volunteers were allowed to stay there free, they just had to have their own sleeping bags. And they asked us for $75 a day, because that's what it cost them for hot water and electricity.

"We took that to the polluters (but) they refused to do anything about it. Then we took it to the Provincial Emergency Program, and they said, 'No way are we going to pay for any accommodations.' So finally Rod Nelson, public relations officer for the Coast Guard, paid it out of funds available to him.

"I don't remember exactly what day it was, but I got so frustrated I phoned the (office of the British Columbia) prime minister, and I also phoned the Canadian military and requested an aid to civil power (asking) that the army step in. I talked to quite a few people in the military, and they said, 'You know, I'm surprised that we haven't been called'."

Webster: "How is that done in Canada? In the United States, the governor or the federal administration must declare an emergency exists before the national guard or federal military can be called on to help in a local situation."

LeBlanc: "In Canada, the Coast Guard can request it, direct. But the Canadian Coast Guard is a civilian organization. When I worked for the Coast Guard I was a civilian in a federal job, not...in a military capacity. The Tofino Lifeboat Station does not do anything without checking with the Rescue Coordination Center (a federal agency) in Victoria. It seems kind of ridiculous to have to ask permission to go out and rescue somebody in the harbor, but...they know where all the resources are at any...time. The lifeboat station doesn't know, but the RCC knows where the helicopters are, where the search and rescue boats are.

"The Coast Guard has a problem with that, because they're taking direction from the military. Ultimately the Coast Guard is going to...be a military organization, but they've dug in their heels, they're not going to go down that easy. So when it comes to the Coast Guard asking the military to do anything, they're admitting weakness. And the Coast Guard will not, absolutely will not admit weakness. They could ask, but they'd have to give up the command position. It became a power struggle, and I'm not just digging this up, because I worked for the Coast Guard. Their excuse was that the military would make more of a mess, but I don't buy that. It seems...a practical way of doing it, because you could enter into an agreement for command and it's an extremely good exercise. But they said they don't want to undermine national security. Are the Russians going to come in here...while we're cleaning up an oil spill? Let's be real. The military people I spoke to...thought it would be quite good. Their Zodiacs could practice landings, and if they had to dig up a bit of beach (to remove tide-buried oil), so what? Long Beach has sand castles built on it by tourists during summer, and it's as pristine as can be. The military would have gone at the drop of a hat, but nobody asked them.

"But as time went on, things just started to fall into place. I finally had it out with the Coast Guard on-scene commander, Colin Hendry, after he took the operation and moved it up to Nootka Sound, totally bypassing us. Because they weren't doing anything here, why have him in charge? I declared him to be stood down and asked for his resignation.

"And...(then finally) the Coast Guard went to the Nuu-Chah-Nulth Tribal Council to contract for help to finish the job.

"Well, we shut down the volunteer coordination center January 26. That was the last day at the firehall, then I continued operating northern groups from home. We sent crews by floatplane, and by wheelplane where they could land on beaches."

Webster: "How long after January 26 did your cleanup continue north of here?"

LeBlanc: "At least another three weeks. But I received oiling reports for a year after that."

Volunteers had responded to the oil on their beaches without expecting to be paid, and they never received a nickel. A claim on their behalf was made part of the U.S. District Court action, which is covered in chapter 11.

10

Common Knowledge

"When there is news of an oil spill, other vessels in the vicinity feel free to pump their bilges and ballast tanks. That is common knowledge," observed Dave LeBlanc, former Canadian Coast Guard employee who coordinated the Tofino, B.C., volunteers.

A contrary opinion comes from Jim Oberlander, spill response team leader for Washington State's Ecology Dept. "I've heard that assertion over the years, but my gut feeling now is that it's kind of outdated. I think times have changed." But workers on Washington beaches found non-*Nestucca*-type oil, he agrees. It could have been accidental spillage. Another example was bunker type oil cleaned up from Protection Island, 70 miles eastward in Juan de Fuca Strait. Although labratory testing identified it as *Nestucca* oil, NOAA's hazmat experts said currents and distance prevented it from reaching there, and Sause declined to pay for cleaning it up. But a maritime author, James Gibbs, has published historic evidence otherwise, as mentioned in chapter 4 (cite 4-4), and it may in actual fact have been from *Nestucca*.

LeBlanc cites a specific incident relative to the *Nestucca*.

His information came from a former crew member on the tanker *Thompson Pass*, hauling Alaskan North Slope crude from Valdez to Long Beach, California, for BP (British Petroleum). LeBlanc's claim is buttressed by the oil industry's "Guide for Selection of Tankers" (10-1). The guide employs an industry rating system using the digit 5 for the finest and most qualified vessels and 1 for the oldest and least dependable ships.

The *Thompson Pass* is a U.S. flag vessel rated at 176,000 dead-weight tons (the quantity of cargo she can carry), built at Avondale Shipyards and registered in Wilmington, Delaware. She was 13 years old in December, 1988, the time of the *Nestucca* spill. Her rating was "1" in the guide volume, which also listed the following significant items:

June 1986--Hull fracture at sea, $30,000 for repairs.

August 1988--Hull fracture at sea, cause unknown, no report on repair.

The former crew member, who had been aboard at the time in question, told LeBlanc the vessel had hauled a full cargo of crude south from Valdez late in 1988, and was directed to deliver it through the Panama Canal to an unspecified destination. Upon arrival, a bottom plate fracture in its hull was reported, and it was refused entry to the canal. The tanker returned to Long Beach for discharge of its cargo, and then was dispatched north to Valdez again, where in January of 1989 it spilled 220 tons of crude (58,667 gallons) in Valdez harbor while being loaded.

What is of particular interest in relation to *Nestucca* is that the *Thompson Pass*, according to the information given LeBlanc, was passing offshore just as the leaking barge was being towed away from Grays Harbor.

But there is more. LeBlanc says a sample was collected during the Vancouver Island cleanup of some oil "different" from the *Nestucca's* and was sent in for testing. The test result was inconclusive, according to information given LeBlanc, allegedly from BP. Different oil also was found on Washington beaches.

Another piece of information that ties in was January news from Oregon beaches of oiled birds coming ashore, at least 74 from Astoria to beyond Lincoln City. At the time the oil was attributed to *Nestucca*, but that oil had moved north, leaving the presumption that the oil off Oregon came from some other source (10-2).

LeBlanc adds one more observation relative to tankers hauling crude from Alaska south, pointing out that the Gulf of Alaska and the North Pacific generally provide some of the world's worst winter storm conditions and most treacherous waters. Deep-laden tankers are regularly swept from stem to stern by monstrous waves that create stresses on aging hulls. One might assume it is only a matter of time until a tired tanker becomes a storm victim, spreading its cargo of crude oil off the Pacific Coast. The *Exxon Valdez* spill in March of 1989 in Prince William Sound released more than 40 times as much oil as did *Nestucca*. And that was not the tanker's full cargo.

TOTAL OIL TRAFFIC ASTOUNDING

There is general public knowledge that oil, both crude and various processed petroleum fuels, is transported both on and around Puget Sound. Most common is the sight of a fuel truck trundling down the public highway.

Less well-known are the huge quantities transported in and out of and among various Puget Sound ports by tankers and barges. In 1989, in response to a request from the States/British Columbia Oil Spill Task Force to the Washington petroleum and maritime industries, a report was prepared by the Institute for Marine Studies at the University of Washington (10-3).

Its figures are astounding. The report says 243 million barrels (more than 10.2 billion gallons) of petroleum were transported on

Puget Sound and the Strait of Juan de Fuca during 1988--186 million barrels by tanker, 57 million barrels by barge. Total petroleum flow in 702 tanker direct transits was 186,031,783 barrels, which is some 7.8 billion gallons. The daily average was 509,676 barrels, 21.4 million gallons. Most of the volume (93 percent) was inbound Alaskan crude being delivered to the refinery ports of Anacortes and Ferndale in northern Puget Sound.

Distribution activity included 1,101 trips carrying 44,010,662 barrels or 1.85 billion gallons. Of that quantity, 3.8 million barrels (8.3 percent) involved Vancouver, B.C. Ferndale shipped more than 6 million barrels out from Puget Sound in 58 trips, and Anacortes shipped 1.78 million barrels out of the sound in 38 trips. Distribution from Anacortes to other Puget Sound points was 5.06 million barrels, and from Ferndale to Puget Sound points, 11.9 million barrels. Those two ports handled more than half the distribution traffic, Anacortes 27.6 percent and Ferndale 24.8 percent.

Seattle handled another 24.9 percent of the traffic, including 217 out-of-Puget Sound trips carrying 1.37 million barrels. Tacoma also shipped 2.3 million barrels in 23 trips outbound from Puget Sound. This outgoing cargo was mostly refined product, in large part gasoline, diesel and aircraft fuel. Bunkering activity, providing fuel to ships visiting Puget Sound, accounted for 433 local trips delivering 4.095 million barrels at Seattle and 383 trips south and north delivering 2.5 million barrels, nearly half the bunkering total. Tacoma bunkering activity accounted for 430 trips delivering 3.2 million barrels, which represented 23.2 percent of that business. Altogether, bunkering represented 10 percent of oil transport on Puget Sound. Refined products accounted for 36 percent of the total, and the remaining 55 percent was crude oil delivered to refineries.

Total steamship (meaning all large powered cargo vessels) traffic for the year 1988, provided by the Marine Exchange, was 3,274, with 572 tankers of which 77 flew foreign flags.

No later figures on total Puget Sound oil transport are available, although the Volpe Report of 1997 (10-4) provides related material in voluminous detail. It lists Puget Sound tonnage total in 1994 of 129,278,326, of which 67,633,000 tons was for Vancouver, B.C., 22,335,514 for Seattle, 17,615,819 for Tacoma and 12,950,108 for Anacortes (mostly oil). Since 1977, according to the report, Alaska North Slope oil carried from Valdez has provided 99 percent of crude oil coming into Puget Sound, citing 282 tanker transits in 1990. Inbound tankers carry an average of 13.5 million gallons of crude oil, and outbound ships an average of 5 million gallons of refined product. Commercial cargo vessels take 250,000 to 2 million gallons of bunker fuel apiece on departure.

A sidelight to tanker traffic on Puget Sound is the observation of an Olympia friend, Bob Benton, retired from a career on NOAA (formerly U.S. Coast & Geodetic Survey) vessels. Navigating a large tanker through Prince William Sound in Alaska is simple and straightforward compared with maneuvering such a vessel through the San Juan Islands' Rosario Strait in northern Puget Sound to reach the oil refineries at Anacortes or Ferndale, says Benton.

Tug escorts are now required under federal law for bulk tankers over 5,000 gross tons east of Port Angeles, in the portions of Haro Strait and the Strait of Georgia under U.S. jurisdiction, in Rosario Strait and elsewhere in Puget Sound (10-5). Only one tug instead of two may be sufficient for double-hull tankers in state waters under regulations recommended by the Office of Marine Safety in May 1993. Additional aids to navigation were proposed in a few locations, inauguration of VHF marine band and cellular telephone communication systems for the coastal stretch from Pacific Beach to Cape Flattery, along with standard qualification criteria and an emergency plan for tug escorts.

Stationing an emergency response tug at Neah Bay has long been a goal for oil spill prevention, since tugs from Port Angeles are not always available and in case of storm conditions could not respond to a situation 70 miles west in the Cape Flattery area or beyond. "It's a no-brainer," asserts Barry Troutman, wildlife biologist with Washington State Fish & Wildlife's spill response section. He knows from experience on NOAA vessels that gale winds blowing easterly in Juan de Fuca Strait make westward navigation impossible, but that a tug based at Neah Bay could reach a ship in distress quickly.

In September of 1999, Washington Gov. Gary Locke called on the federal government to station a tug at Neah Bay (10-6). The cost is estimated at $250,000 to $500,000 a month, Governor Locke said. The Coast Guard was studying the problem.

Finally, on Oct. 14, 1999, the U.S. Department of Transportation found $250,000 to help station a tug at Neah Bay (10-7)

That same day, by happenstance, two laden tankers bound for Puget Sound were disabled. The *Angelo D'Amato* was in Juan de Fuca Strait approaching Port Angeles to pick up a pilot, as required for inbound vessels, when it lost power. Tugs nearby helped the ship, carrying 25 million gallons of crude oil, to a safe moorage. A second tanker, the *New Endeavor*, loaded with 11.3 million gallons of jet fuel destined for Seattle-Tacoma International Airport, lost its steering well offshore; fortunately, the weather was not stormy and tug assistance could reach it.

That DOT commitment was a start, and Governor Locke was able to promise an additional $100,000. John Neel, a senior officer of State Ecology, said in early November that accumulated interest on *Tenyo Maru* settlement funds, in excess of $400,000, could be added to the pot for a total of more than $750,000 available for the coming winter storm season. That didn't put the tug in place, but made it conceivable pending actual arrangements.

Another 1999 development was the federal court suit brought by Intertanko, a Norway-based organization representing three-quarters of the world's oil tank vessels, contending that Washington State's tight regulation of tankers visiting Puget Sound should be invalidated as violating federal law and making more difficult international tanker operation. The Ninth U.S. District Court had supported the state's right to enforce tighter safety regulations. Intertanko, however, appealed to the U.S. Supreme Court, and gained the support of the U.S. Attorney General and several other countries including Canada. After oral argument in December, the high court was expected to issue its decision sometime in 2000. Washington State was supported in its strict requirements by 29 other states, contending that tougher laws are needed to help prevent catastrophic spills in restricted waters such as Puget Sound, traveled by hundreds of tankers, some of them renegade foreign vessels with multiple registrations, non-English speaking crews and hard-to-trace owners (10-8).

However, this is getting ahead of the *Nestucca* story, and we'll back off to the 1990 U.S. District Court trial in Portland.

11

U.S. District Court Decides *Nestucca*

Trial of the *Nestucca* case took a week during December, 1990, before Judge James A. Reddin in United States District Court, Portland, Oregon (11-1). It involved only the question of liability, as damages had been separated by the court for later determination following the more basic decision as to legal responsibility for the accidental oil spill off Grays Harbor two years before.

As the session convened in the high-ceilinged, fourth-floor courtroom in downtown Portland, Judge Reddin was advised that the United States and the State of Washington had agreed to a $3.35-million settlement with Sause Brothers Ocean Towing, subject to approval by the U.S. Departments of Justice and the Interior in Washington, D.C. That left Canadian interests as the major remaining participants in the case.

But that was a side issue to the District Court trial, preparation for which had built a two-foot pile of papers filed by attorneys for the various parties to the action. Legal depositions--pre-trial testimony given under oath--accounted for about 1,200 pages alone. Depositions from the *Ocean Service* skipper, crew members and other Sause people involved helped provide details for the account of events in this book.

The settlement left the state and federal attorneys in mere observer status during the trial. Lead attorneys for the ensuing case were Guy Stephenson of the Portland law firm Schwabe, Williamson & Wyatt for the towing company, and Douglas Fryer of the Seattle law firm Mikkelborg, Broz, Wells, Fryer & Yates for Canada. A number of attorneys representing other parties also were present, some of them taking an active part.

Fryer's opening statement for Canada asserted that Sause improperly outfitted its tug and barge for the December 1988 voyage from Ferndale, Washington, to Portland with a stop at Grays Harbor. Condition of the towing cable was such, he argued, that rust and deterioration from use, along with wear on the (steel) chafing plates (on posts and towing pins) and fairleads, perhaps a hangup of the catenary (see again fig. 3, page 19) on the ocean floor, combined with lack of required inspections and maintenance, resulted in the cable parting under moderate sea conditions.

Stephenson, for Sause Brothers, argued in his opening statement for limitation of the action based on discretion of the master of the vessel under the doctrine of perils of the sea. The remainder of the towing cable, some 900 feet of which was left on the tug's winch after it parted, showed 414,000-pound static breaking strength. The tug, with 3,800 horsepower, was unable to exert enough pull to break the cable, he said. There are, he asserted, no standards covering such cables nor their maintenance. Cable strength had been reduced by some 20 percent when towing pins at the tug's stern were lowered preparatory to crossing the Grays Harbor bar, so that the cable could swing in a larger arc. He claimed that the skipper's actions were prudent in the circumstances, that the Orville hook did its job and that reasonable actions should prevail in the court's decision.

Reason prevailed, but not Mr. Stephenson's argued points.

Most of the ensuing testimony related to the towing cable and events of that fateful night of December 22, 1988. Witnesses included Captain George K. (Kurt) Greiner of Ridgefield, Washington, retired senior investigator for the Seattle Coast Guard District, with service also in Alaska and New York, an expert in towing cases, collisions and sinkings, called into the case by the State of Washington.

Others were Alard Berend Ages, hydraulic research engineer; Paul J. Irish, chemist, metallurgist with Northwest Testing Laboratory; Wilbert Lucht, wire rope specialist and metallurgical engineer; Capt. John S. Blank III, retired mariner and author of a book on towing who had skippered the *Ocean Service* under a previous owner; and William I. Millwee, retired Navy officer and senior vice president of a diving and salvage firm.

For the Sause Brothers side of the case, testifying were George A. Barnum, another retired Coast Guard marine investigator; Ian McG. Heslop, metallurgical engineer from Vancouver, B.C.; Cmdr. Douglas R. Halsley, (ret.) USCG; Capt. William P. Figari of Crowley Maritime; Gary Rickey, mate, and Capt. Charles E. May, skipper of the *Ocean Service*; Capt. Pat Kellis, Coast Guard retiree and senior master with Sause; and finally, Capt. Jerry White, port captain for Sause.

Mr. Lucht, for example, told how he had taken pieces of the fractured towing cable for laboratory examination, including photos at 10x to show the nature of the rupture and others up to 5,000x magnification to show changes in the wire. He also pulled some of the pieces to determine stress fracture points and other features, and tested new wire of the same type and manufacture for comparison.

Some of Lucht's pictures showed, he said, drastic changes caused by angular pressure and strain such as will result from passing the cable ("rope" as in "wire rope" was the term used most of the time by these experts rather than "cable") over or around an object causing a sharp bend where load was applied. There is, he said, no way to determine the remaining strength in such a piece of cable; and in addition, breaking is a function of how far the wire will stretch before breaking. Corrosion, which was present in the samples, also reduces stretch, he added.

Shown pictures of the tug's stern towing pins, he noted deep corrugations in them. "This is a cardinal sin," said Lucht, explaining that sharp, high points induce wear and put more pounds per square inch on the wire.

"Is bending around a small diameter bad for wire rope?" he was asked. "Yes, many tests confirm it," replied Lucht. "Any rope man even looked at that (towing cable), he would throw it in the scrap can."

The *Ocean Service* cable had about half its dynamic breaking strength, Lucht believed.

He also testified relative to the need for daily inspection of towing cable, in detail at close range, mainly for broken wires. The basic standard, he said, is that three broken strands per lay (length of one strand of cable once completely around the overall cable) or six total per length, then that cable should not be used. Inspecting a piece of the broken cable in court, Lucht found 11 broken wires and many more abraded wires, all of which he said pre-existed the Dec. 22 event.

Advised that the tug crew members said they inspected the towing cable from 8 to 15 feet away as it was brought in or out, Lucht said, "It is impossible to inspect at that distance."

Captain Blank was another who testified that the cable used to tow *Nestucca* was unsuitable. "I was shocked at its appearance. It showed great use and great wear. Absolutely, it was not suitable for use." He also said the stern roller on the *Ocean Service* was in bad shape. "It would cause abrasions, great wear on the wire. It's supposed to be absolutely smooth." As to the fairlead rollers, "I can't believe they would be used. All kinds of wear and grooves. The engine (of the tug) is most important, but the towing gear is second most important."

As to entering West Coast "doghole" harbors, Captain Blank explained that tug skippers try to "float" their tow across the bar. Asked if the towline can be kept off the bottom during such a transit, he said, "No. The surge chain, bridle and pigtail weigh from 11 to 13 tons."

In contrast to the foregoing testimony on behalf of Canada, British Columbia and the other parties in the case, Ian Heslop, testifying as metallurgical expert on behalf of Sause, said that his inspection of 900 feet of the remnant towing wire turned up but eight broken wires, which he said was well within accepted practice. Although he agreed that both corrosion and abrasion were visible, in his opinion failure was not from those causes.

Asked whether the *Nestucca* wire was within the safety factor relative to breaking strength, he answered, "Yes, it is more than the recommended safety factor."

After having seen Mr. Lucht's reports relative to bending wire around wear plates such as used on the Sause tug, Heslop told of making test jigs to simulate wear plate conditions. The strength reduction, he said, was 31 percent around the jig as compared with a straight pull. Using new 2 1/4-inch cable he recorded a 23-percent reduction, leading him to the conclusion that the *Ocean Service's* cable strength reduction may have been as little as 10 percent.

Heslop believed the towline must have been snagged on the bottom, that something had to cause the difference in the force required to break the cable. He cited as evidence the formation of martensite at the fracture site, a phenomenon requiring high temperature typical of friction violence.

He also had noted damage to the cable 32 inches from the fracture site, and upon measuring the tug discovered that from the starboard side of the level wind to the wear plate is exactly 32 inches.

Commander Halsley, working as a consultant to the Coast Guard on a committee reviewing tug/barge standards, confirmed the lack of standards under extant USCG regulations, but said they were being developed and would be in effect within the next few months.

Captain Figari, with a degree in foreign trade, an ocean master's license and 45 years experience as a ship master with Moore-McCormack and towing for Crowley, affirmed the Sause policy of depending on the tug crews to inspect towing wires. He also said Sause Brothers Ocean Towing is recognized as expert in the industry, and that reliance on the Orville hook rather than on insurance wires was "prudent, in light of their experience." In his opinion, Captain May was acting reasonably in his situation, and "did everything I would expect one of my people to do."

Asked what, other than a wreck on the bottom, could have caused the cable to break, Captain Figari responded: "Perhaps a boulder on the shifting sandy bottom. Sometimes cables can fail in calm weather or crossing a bar, where things can be uncovered. My people have hung up wires when nothing was known to be there."

Captain May's personal testimony on Friday, Dec. 7, the final day of the trial, included the information that he had 20 years with Sause, covering 644,000 miles of towing and 16 years as a tug skipper, with four years on the *Ocean Service.*

He described procedures and some details of equipment carried aboard the tug, then testified as to events prior to the cable break.

"When I start taking in wire, I go back to the cable control and slow the boat to about 350 r.p.m., but still moving, from idle to 500 r.p.m. during retrieval. I saw no bad spots on the wire. Then I set the air brake, disengaged the engine from the winch, left the engine turning at about 600 to 650 r.p.m. and went back to the wheelhouse.

"Picked up a few more turns (r.p.m.) in the wheelhouse, wanted to move a little further south on the range, and wanted to check the progress of an outbound ship. Wanted to come in south of the Point Chehalis Range because of southerly wind. I was pretty much on range, or a little bit south or southeast, and I turned a bit more so, to approach the bar.

"The towing pins had been down since before retrieval of the wire...to allow it to angle more across the...quarter, up against the...frame. I had increased to probably 700 to 750 r.p.m., which I can feel by the throttle, but did not pay much attention to the actual speed.

"The wire broke at about 2345 (11:45 p.m.), but I don't know the exact time. I was pretty busy. Heard a noise, a loud pop or bang. Turned into the trough (of the swells), and we were rolling around quite a bit.

"...When the wire parted, the barge continued drifting in an easterly direction. I had buoy 5 for reference. The tug was inside buoy 3, and the barge was near buoy 3 in 45 to 50 feet of water. The barge changed its direction, turned on the wire (which was dragging on the bottom) going bow into the seas, and started drifting north. I could see breaking white water in the vicinity of buoy 5.

"I ordered the crew (Rickey and Bartley) to break out the nylon for the Orville hook. They have to pull it out of its box behind the stack, flake it out back and forth on the fiddley (the deck atop the aft cabin) ready for use."

He was asked by counsel, "Was it your intention to deploy it?" Captain May answered, "Yes, but I changed my mind. At the rate the barge was drifting, I figured I wouldn't have time. So I sent the two men aboard to pass a heaving line and pull a hawser aboard the barge

"As we approached the barge, backed down toward it, the two men were on the poop deck (the stern of the tug), and I was at the aft controls, with three floodlights on, so we had excellent visibility. I had the tug angled 30 to 45 degrees away from the barge. You need the angle to help your getaway. We had to make two attempts, and after I put the men aboard on the second try, I became aware that the boat was not acting right. That's when I found the steering jammed. ...Then I backed down close to the barge again and told them to stay in the pumphouse. Conditions were changing, the wind was building. I didn't want to worry about the men on the barge.

"At 0030 (12:30 a.m.) I radioed that we had two men on the barge (reading notation from the wheelhouse log). At 0105 sent a message that I had the barge in tow after using the Orville hook. That time included a steering check in the engine room, hydraulic levels for the steering. I first thought we'd have no steering at all, but at full power managed to break the rudder loose."

This much of Captain May's courtroom testimony compares favorably with his deposition, differing only in minor details of his recollection.

The trial wound down to its afternoon end, concluding with Judge Reddin's observations: "The main issue here is the wire rope, its condition. The snag theory seems to be speculation. Had the tug been going 3 to 3 1/2 knots, a snag would have brought it to a complete halt. That's a weak reed. What did happen, no witness was able to tell us. And...there is evidence on both sides. Test of sections of the rope after the fact showed good strength. It seems possible that defects were not seen, or were seen and not appreciated, or were even invisible. As to the crew, what is significant is were those crewmen capable of doing their tasks, or if not, was the company negligent, or is there privity? Is it a case of he knew or should have known? If there is no negligence, the result is exoneration. If there is negligence, is it knowing and willful, in which situation there is no limit to liability."

Those off-the-cuff comments perhaps implied but failed to telegraph the judge's written opinion, issued January 24, 1991. In 22 pages the jurist outlined the details of the case, laid out applicable law and explained his ruling. Excluding most of the citations, its highlights follow:

"For the reasons which follow, I find SBOT (Sause Brothers Ocean Towing) cannot be exonerated from or limit their liability, pursuant to 46 USC App. par. 183(a), for the damages incurred by the claimants in this action. ...(This) act was...(adopted) in the United States in 1851...after laws of...seafaring nations...limiting a shipowner's liability to the value of the vessel and its cargo. Claimants have the initial burden of proving negligence or unseaworthiness; once established, the burden then shifts to the vessel's owner to prove lack of privity or knowledge.

"The determination...is left to a two-step analysis by the court: (1) what, if any, acts of negligence...caused the accident, and (2) whether the vessel owner had knowledge or privity of those...acts...or conditions.... To prevail, claimants must prove negligent acts caused the casualty and...were within the privity and knowledge of SBOT.

"The winter voyage...could expect heavy seas and four bar crossings (sic: only three, ending at Portland) at Grays Harbor and the Columbia River. The oil carried by the barge presented a significant risk to the environment if a spill were to occur.

"...A tow wire should...be able to withstand the size of the load as well as the dynamic forces placed upon it by the rough seas. At the time this...wire parted, conditions were moderate and the tug was towing the barge (at)...three to four knots while maneuvering for the bar crossing. The load upon the tow wire was substantially lower than the 487,000-pound tensile strength it was rated to withstand. The portion...that was saved and examined showed that it was corroded and abraded prior to its failure....

"Claimants argue that SBOT was negligent in their policy not to lubricate tow wires as recommended by the manufacturer.... It appears that while lubrication...is not required, it helps slow...corrosion. ...According to one expert (witness), when the wires slide against each other, absent lubricant, this weakens the wire and contributes to lower load capacities.

"The tow wire...abrasion could have been caused by a number of factors. ...SBOT admits that...dragging the wire on the ocean floor...could abrade the wire. While...(this) cannot be avoided, such conditions would necessitate careful inspection of the wire as it is rewound on the drum. Evidence also showed that the (fairlead) rollers on the *Ocean Service* were very worn and could have caused... abrasions.... Testimony...suggested that the chafing plates...were worn. These...were replaced...and discarded before they could be examined or even photographed. Based on the testimony and the worn condition of the rollers, I will infer that the chafing plates were in a similar...condition and contributed to the wear upon the tow wire....

"SBOT's normal inspection procedures entailed the master of the tug viewing the wire as it was drawn in or let out from (his position at the controls on the) boat deck and the chief engineer...from the fairlead...controls. Both...were from six to ten feet from the wire.... The wire was drawn in at...one to three miles per hour. At this distance and speed,...inspection is inadequate to detect breaks,...abrasion or corrosion....

"SBOT contends it required its masters to document wire inspections in the log book. No such entry occurred in the *Ocean Service* log book in 1988. Trip reports are also to be submitted (yet none had been)...between May, 1988 and the date of the casualty. SBOT also required the tug masters to utilize a marlin spike to separate the tow wire and inspect its core. Captain May testified that he had never used (it)...because he believed it damaged the wire.

"...The tow wire was upended (reversed) and a new towing "D" poured December 11, 1988 (just 11 days prior to the casualty). The crew stood at the stern of the tug and washed...the wire with a fire hose as it came aboard. Captain May testified that the wire was inspected at that time. However, (he) admitted that the upending... was not completed until after dusk and the *Ocean Service's* lights were used to finish the procedure. (This) would leave ample opportunity to overlook signs of abrasion and corrosion that would indicate dangerous weakening of the tow wire. These are not reasonably acceptable methods of inspection.

"The preponderance of all the evidence before the court demonstrates that the tow wire was defective and SBOT was negligent in...maintenance and inspection....

"Claimants contend that the doctrine of *res ipsa loquitur* (Latin: the thing speaks for itself) applies in this instance.... (However) the owner of a tug is not an insurer (and) loss of a tow does not raise the presumption of fault (and) further, the parting of a tow wire does not give rise to the presumption of negligence.... Therefore, the doctrine...is inapplicable in this action."

Two lesser violations are noted, first, that the statutory 12-hour watch limit had been exceeded by Captain May when he stayed on duty to follow company practice for the master to take the tug and barge over the Grays Harbor bar. A second technical violation, of the requirement that each of the tug's crewmen be qualified as an able seaman, left a burden on the towing firm to prove it could not have contributed to the accident. But, the court pointed out, "none of the crew, except for Captain May, had ever deployed the Orville hook or practiced assembling it. Once the line was flaked out, Captain May decided he had insufficient time to retrieve the runaway barge with the hook. Since (it) was not equipped with an insurance wire, Captain May was left with only one option: putting two men aboard the barge and heaving an emergency towline back to the tug. Putting men aboard a barge is the most dangerous method....

"An insurance wire consists of a spare towing wire pre-rigged for rapid deployment in an emergency.... The wire is fastened to the barge's bow, down the side and coiled on the stern...connected to a floating line and buoy that trails behind the barge.... Insurance wires are not without their problems, e.g. if the buoy is washed aboard the barge in heavy seas, or if the line becomes tangled, they are unusable....

"The ultra hazardous nature of the cargo transported, coupled with the inexperience of the crew, contributed greatly to this casualty. I find SBOT negligent.... The burden of showing lack of privity and knowledge of the negligence that caused the spill now shifts to SBOT. (Here Judge Reddin quotes a Federal Appeals Court case, Waterman, 414 F. 2d 738) 'The measure in such cases is not what the owner knows, but what he is charged with finding out.'

"...For these reasons I find SBOT had privity and knowledge of the negligent acts that caused the casualty....(and) is ineligible for limitation of liability...."

Judge Reddin's decision resulted in payment of more than $10 million as Sause settled claims by Canada, its Province of British Columbia and the Nuu-chah-nulth Tribal Council of Vancouver Island aboriginal groups. Following a court-referreed conference in the spring of 1992, the federal government of Canada and the British Columbia Ministry of Environment, Lands and Parks agreed to a joint settlement for $8.7 million to cover environmental damages from the 1988 spill. A federal/provincial environmental trust fund was to receive $4.3 million for habitat restoration and coastal protection projects administered through the Canada/B.C. Wildlife Agreement. Another $4.38 million was to repay federal cleanup expenses, in addition to $380,000 previously agreed upon for provincial cleanup costs. That made the total $9.06 million.

Negotiations with representatives of the Nuu-chah-nulth Tribal Council produced a settlement under which Sause paid $1,205,000. Richard Watts, an officer at the tribal offices in Port Alberni, said that after paying all individual claims from tribal members they hoped to wind up with more than half that amount in a trust fund reserved for future environmental damage situations.

"We were there first," Watts said of tribal members who responded to *Nestucca* oil on their home island's shoreline. Natives of the 14 tribes that make up the council were reimbursed for oiled clothing and gear, for boat use and other identifiable expenses during the cleanup. Tribal members, he noted, paid attention to food resources such as clam beds and saltwater fish spawning areas. "Those are our heritage." Some tribal members, Watts said, were reimbursed for work with non-tribal volunteer beach cleaners directed from Tofino.

Non-Indian volunteer cleanup workers were never compensated. LeBlanc, their leader and spokesman, said they filed a claim in the U.S. District Court action for $3.5 million, based on their work and expenses. After settlement of the major claims by other parties, their class action could not be maintained in such a maritime liability proceeding.

The hope that a separate action could be filed in Canadian courts did not materialize, LeBlanc said in 1999. He recalled that in frustration he made a telephone call to Dale Sause at the company's Coos Bay headquarters to tell him, "We have unfinished business," and to propose that they take a walk on the beach to talk together in hope of reaching agreement on a settlement.

In response, said LeBlanc, Sause "accused me of costing him millions of dollars" in expenses and payments, presumably because he and other volunteers began and organized the spill cleanup in British Columbia.

Sause, he said, spoke of $50 million. Known expenses as a result of the quarter-million gallon spill were some $2 million paid for beach cleanup, plus the financial settlements following court action. Those were $3.35 million for Washington State and the U.S. federal government, $9.06 million paid to settle the Canadian and British Columbia claims, and $1.2 million paid to the Nuu-cha-nulth. The total of these figures is some $15.615 million, but of course the towing company had legal fees and additional costs.

The volunteer worker spokesman was particularly rankled by a Sause allegation that the Canadians made "frivolous claims" for "insignificant" efforts when the oil spilled off Grays Harbor spread north to Vancouver Island and came ashore on their national park and other beaches. "Next time," says LeBlanc, asserting that civilian Canadians are ready and waiting to make up for being left out of the *Nestucca* settlements.

First People Meet Second Assault

Participation by the Makah, Quinault, Nuu-Chah-Nulth and other native tribal folk in the *Nestucca* events has been mentioned previously, but these "first people"--as they correctly refer to themselves--have had to respond more than once since 1988-89. They have not been idle, and highlights of their story merit inclusion in related events.

A major happening since *Nestucca* was the *Tenyo Maru* oil spill July 22, 1991, off the Vancouver Island coast. The *Tenyo Maru*, a Japanese fishing vessel, was rammed and sunk by the Chinese freighter *Tuo Hai*. For more than a month, the sunken vessel released fuel (various oils, including diesel, lubricants and fish oil) from her resting place 550 feet underwater. A 40-mile oil slick fouled beaches from Vancouver Island south to Oregon. The 4,000 birds found dead were an unknown fraction of the actual fatalities, and most of those found were murres.

On the second anniversary of the spill, July 22, 1993, the Washington State Department of Ecology assessed a $1 million fine against the China Ocean Shipping Co., owner of the *Tuo Hai* (12-1). Evidence indicated the *Tuo Hai* was traveling at high speed in poor visibility through an area swarming with fishing vessels, and did not heed radioed warnings from the Coast Guard, perhaps because the ship's people did not understand English. The shipping company was required to post a $15 million bond in Canada before its ship was allowed to sail from Vancouver, and Canada spent about $3 million on the cleanup.

Legal claims against the owners and insurors of the two vessels by the U.S. Department of Justice, State of Washington and the Makah Tribe were settled in late 1994 with a $9 million payment intended to cover civil penalties, cleanup costs and natural resource damage claims (12-2).

Because of the Makah people's uniquely sensitive position relative to coastal oil spills and active involvement in coping with two major successive events, we visited Neah Bay to interview key people. Vince Cooke, the tribal hazardous materials (hazmat) coordinator, also carried responsibility for protecting the traditional tribal village locations and the cultural heritage involving natural food supplies.

Cooke told how tribal workers were trained to handle beach cleanup during the aftermath of *Tenyo Maru*. "We employed 112 Makah people, all OSHA certified with eight hours of training, and they worked both along the ocean coast and the strait side. As we sent these groups out, our main reason for being involved was the cultural areas, protecting them from being disturbed by out-of-town contractors. We would send our own teams to clean where we knew there were sacred sites. That's always been very important to us."

During cleanup after earlier oil spills, bulldozers had plowed through known ancient village sites, he said. For lack of communication and involvement with tribal people, work proceeded with no awareness of the cultural damage involved.

"There was a stickler down at Cannonball, on the Ozette Reservation. The island is a culturally sacred site, with some burials there, and petroglyphs (rock pictures). The rocky cobble (between the island and the mainland beach) was impacted...by bunker-C fuel. We cleaned it up as much as we could, then the natural resource damage assessment team re-evaluated it, and we refused to sign off (approve) because to us it was not clean enough and the situation was disrespectful to our cultural area, even though the Coast Guard cleanup supervisor decided it was good enough. We contacted the office staff of one of our congressional representatives, and about a week later we were informed the Coast Guard would send a team back to clean it properly. And they did.

"It helped us create a respectful relationship with the Coast Guard," explained Cooke, "to show them we have connections outside, and when there's a problem like that, we take it very seriously."

Another aspect Cooke was working on was getting the Makah's team of trained and qualified oil spill cleanup workers under contract with the Coast Guard to respond to spills elsewhere, similar to the manner in which tribal firefighting teams assist the U.S. Forest Service. Unemployment at Neah Bay runs about 75 percent, so a cleanup contract between the Makah Tribal Council and the Coast Guard could help.

Cooke outlined tribal history from the time of the 1855 treaty with the U.S. Government, when there were five main villages. There were whale hunters, river and salmon fishermen, seal hunters and hunter-gatherers, all sharing resources from the 44 square miles of the reservation, plus the mile-square Ozette reservation. All the people shared the Kwidishda-akh language (12-3).

"Then in the early 1900s," said Cooke, "The government came and said, 'You're going to have to abandon all those other villages and move to Neah Bay, because that's where the Bureau of Indian Affairs school is.' So everybody didn't want to abandon the villages, but we were forced to. There is a school here, that's how we became one village. Our tribal jurisdiction goes a little past Ozette, and our 'usual and accustomed fishing area' goes out 40 miles. Our salmon rights go clear up (east) to Sekiu, close to Port Angeles. But we do work with the other tribes on salmon issues."

Cooke, in his sixth year working for the tribe, had started fresh out of college, and was involved in technical developments including the modern museum which features exhibits from the Ozette archeological excavations of 500-year-old homes buried by mudslides. It's been a learning process, reinforced by Cooke's respect for his grandmother, who still depends on native food sources (as do other tribal elders). A preliminary award for damages sustained by the tribe was distributed among them.

Since our visit, Cooke has accepted a new position with the Quileute tribe at LaPush, some 32 miles south of Neah Bay, and is working to train and qualify a crew of tribal workers there for oil spill response, similar to the team from the Makah/Kwidishda-akh.

Denise Daley, tribal fisheries biologist at Neah Bay, also met with us, and brought with her Bobby Rose, tribal bird rescue worker who was featured in our chapter six account of 1989 bird recovery on Makah Reservation beaches after the *Nestucca* spill. She did the same after *Tenyo Maru*.

Rose and Daley, as they explained, have developed a close friendship since Daley started work for the tribe after earning her fisheries degree at Sheldon Jackson College in Sitka.

An intriguing contribution from Rose was the recollection of her grandfather telling how, as a child, he found oiled birds on the beach nearly 90 years ago, and was advised by his mother to wash them in fresh water. The mystery of how those birds were oiled was not solved by researching Gibbs' *Shipwrecks off Juan de Fuca* (12-4), but it might have been during 1918 when salvagers worked seven months and finally freed the Japanese passenger ship *Canada Maru* from a reef off Cape Flattery. Oil may have been employed to quiet rough water, an occasional practice, with the unfortunate result that seabirds were incapacitated.

There is a special quality in the account Rose gave in her own words, so after I asked how soon following the *Tenyo Maru* sinking did oil appear on the Makah beaches, Rose takes over:

"Very first day. I live on the beach. There's a rock they called Tsawes, means round thing on the beach, nowdays they call it strawberry rock, it's where all the birds from the ocean that are hurt, the first place they go. Mukkaw Bay, on the ocean side, the second river going south, Sooes (pronounced Soo'-yez). Me and my kids went down there, because all my kids worked the first spill (*Nestucca*), three of them worked the second spill.

"So then people started bringing more birds, and I held them at my house until I had a couple boxes full of birds. And I was trying to get help through the tribal council, but because oil spills were very new for the council, for this village, I had to bring birds up to the council meeting and beg for space for primary care, because by then I was outgrowing my garage. So then we finally got permission for the fire station at the old airbase, and then it grew from the first room to the second room, to the third room; and we had the morgue room, so we had the whole building. But out there we had no running water, we had no toilet. We had electricity, but we had no meals.

"I worked four and a half or five days with no sleep, no place to wash your hands or anything, and we had no food. The first thing we ate was we begged his lunch off a guy walking by to work. And we shared that with half a dozen, between my kids and school kids, because I'm a teacher, them and one other adult. You'd be surprised how far a plum can go when everybody's hungry.

"Operated by myself, never saw anybody (else), I believe three, four days. And then the state came in. They brought their portable showers and their trailer kitchen. And my group, that walked the beach, that ran the primary care and the morgue unit, were not allowed to eat with them or use any of their facilities. That changed after about a week, then we could go eat."

One hears in the Bobby Rose account an honest spontaneity not to be denigrated nor equivocated with. I won't try to improve it beyond necessary clarification.

Daley brought up the problem of "The clash between the reservation and when the state comes in. They try and assert their authority and go by their rules."

So Bobby Rose rejoined the conversation. "They say under (age) 18, couldn't work. I had a 12-year-old, an 8-year-old and a 6-year-old. The state told me to get them out of there (12-5), but those were the ones that were there every day with me before the primary care unit. Those kids, when I'm dead, they're all tribal members, are the ones that are going to take care of the beaches. Right now at home I have two kids that can run a primary care center.

"The state, we fought and fought,...(they) wouldn't leave me alone. The first, probably five days, when all the governments came, then you had the Feds fighting, you had the states fighting, Oregon, Washington, Alaska and you had Canada. Everybody was trying to say, 'It's your fault.' They were all worried about the money. That was the big war. You couldn't get any supplies, you couldn't get any tubing gear for the birds, you couldn't get any pedialite (for feeding birds). I bummed from the store, I bummed from my friends that have jobs, I bummed from my mother, I bummed from my father to get money to do it. It was so poor an operation we set up a gallon jug that tourists who couldn't walk the beaches, because they were closed, they'd shove (in) five bucks, ten bucks or they'd bring us doughnuts...or fried chicken--God forbid, that was horrible. Nobody would eat it...all those legs in a bucket; so I waited until late at night when everybody was really hungry, and passed them out.

"It was too much war zoning. The critters were the last. This is my opinion, and I'm a tribal member. The tribe was just like the other governments, because they worried, 'who's going to pay for this?'

"A lot of people in the village, they were laughing at me for doing the birds, or the sea otters or the eagles, or going to check on whales, and they said, 'That's stupid. For every bird that you clean up there'll be a million next year.' Well, that wasn't so, but most people didn't see the critters until they saw a price tag on them...after the negotiations, and that's what I don't understand. When you can say each common murre is worth $49, then people start to change their tune, when you've got 4 or 5 or 6,000 dead birds."

And Bobby Rose is so absolutely right. The settlement with the tribe was $5.5 million.

I asked Rose and Daley if the problem with the state over jurisdiction and responsibility on reservation land has been solved, and Denise responded.

"It's getting better. I met with Sara LaBorde (of the state Fish and Wildlife rescue and response program), and we had some minor conflict (on) reservation jurisdiction, so I (told) her, 'Remember you're not from here.' Once you get that settled, it comes down to...the need for the tribe to do a well-rounded response (plan and procedure). Who do we want coming in to respond to an oil spill. The state always will claim that it's in state waters, but then I remind them, to get to those state waters you have to get on our land.

"In my opinion," continued Daley, "I want somebody non-profit. International Bird Rescue...(survives) by payment for the damages, that's what keeps them in business. I like them because, one, they're non-profit and, two, there's no politics. They're here for the birds, and they know it's a reservation, it's not state land."

In relation to Tatoosh Island (see fig. 1, page 7), which was hit hard by the *Tenyo Maru* spill, Daley bluntly told both U.S. Fish & Wildlife and the state, "You have no jurisdiction here without our permission. The Makah Tribe got its environment back from the U.S. Congress in 1988, and we're not going to permit anybody on that land without our okay." Finally, she concluded, officials agreed to work with her as tribal representative instead of going ahead in the former bulldozer fashion.

Rose said a similar problem has existed between her and tribal authorities, especially relating to the detailed records she maintained covering every bird she worked with--records that provided the proof on which payments for natural resource damages were paid. "As soon as they found out it involves money, they've got a tribal attorney standing at my door demanding my paperwork. To this day I haven't got that book back." She also has not been provided any funds from the tribe to acquire supplies for a future spill, Rose said.

Of course the tribe has other priorities, many demands for real needs typical of reservations with few work opportunities and high unemployment. But Rose is left understandably bitter. "It's real frustrating, and I try not to think about it. As a tribal member, I was totally bummed out, got no help, no supplies. You know how I got my first body bag? The...store gave me an apples and oranges bag because there's no money; and...there's still no money, so I have to invent everything... that has to do with critters..., and that's the way the next spill's going to be."

One thing that gives Rose hope is the understanding that has developed between her and Daley, who agrees to work on help and support for bird rescue and cleaning with the next oil spill. Denise is Makah, as is Rose, but Daley's official status with tribal government carries more influential clout.

The women talked also of "popcorn food," examples of which include black turbots, periwinkles and other small shellfish. Daley recalled, "My grandfather took me out and said, 'Denise, this and this and this is stuff you can eat.' But I noticed, (after) leaving Sitka and going out to our beaches here, those 'popcorn shellfish' are just gone all of a sudden. Now where do I go for them? And I don't know whether it's due to two oil spills hitting successively, or what's going on. No population studies have been done.

"Non-native people focus on birds in an oil spill," Daley continued, "because they float up on shore. But that is tunnel vision.

Education needs to broaden the vision of people in understanding the damage caused by an oil spill or other pollution and its effect on the culture of native tribal people.

"Focus on a beach, and here at Neah Bay we know that a square foot of a certain type beach will hold maybe a hundred black turbot. And if the whole beach (population) is wiped out, maybe 10,000 square feet, perhaps we could get a penny for each turbot to get that beach population restored."

As Rose expressed it, "There's a difference between a natural resource and a cultural resource. To us, as native people on our own land, everything is a cultural resource. We have more uses for a clam than an off-reservation person who maybe would eat the clam. We use the shell for something. That's a major problem. The outside world looks only at natural resources, and we're looking at all the cultural uses for these resources."

Numerous other events were cited, examples of being refused permission to take cultural items for tribal use. Once it was puffin beaks from dead birds for tribal employment in creating traditional decorations. Another was when a dead whale washed ashore on the reservation, and tribal members were refused permission to salvage the meat, after which it was destroyed with dynamite. U.S. Fish and Wildlife officers probably were involved in both incidents.

Said Daley: "I think it'll never be solved. It's just going to be constant conflict, who has jurisdiction." And Rose added, "If there was an oil spill tomorrow, we'd go through the same thing."

In relation to the state's geographic response planning maps (described in Chapter 15), Rose explained that tribal folk know the locations of coastal villages which existed hundreds or thousands of years ago, but don't want them designated on a map. "It's like telling bobotklid (non-tribal people) 'come, dig here' (for things our ancestors left behind)."(12-6)

Relative to the "popcorn food" intertidal life discussed by Daley and Rose, Cooke said the Makah tribe is establishing transects in areas that might be impacted during future spills, and measuring populations of the "little critters" on those beaches. One biologist is working on that contract, and another spends her summers on Tatoosh Island collecting bird data.

Tradition and culture from a long past continue to lead the Kwidishda-akh and other first people into the future, but today they are harnessing technology and science to protect and continue their way of life.

13

Sause Chooses Self-Improvement

Periodically since the 1988 *Nestucca* events I have attempted to establish communication with Sause Brothers Ocean Towing at Coos Bay, with little success. In 1989, after I had received a copy of the Coast Guard report on the incident prior to it being officially released (because, as a Coast Guard officer said, I was engaged in a legitimate project), a Sause official called to ask if I could provide them a copy. In response, I quoted for them key portions of the report, but did not feel right about making a complete copy. Two other times, I spoke with Rick Kimberley by phone, but have not met him.

In the spring on 1993, having heard from two knowledgeable persons of the advances and improvements Sause had made in preparedness and training for possible future oil spills, I drove down to Coos Bay while we were visiting family in Eugene, Oregon.

Dale Sause, president, received me in his office at the firm's headquarters on the waterfront.

Civil but cool, Sause acknowledged his non-response to my repeated overtures, explaining that they would prefer to have people forget about *Nestucca*. Without saying it in so many words, it was obvious that he would prefer not to have a book published commemorating the event.

My request to Sause was for information relative to their spill-prevention program. He declined, saying "We see no need to inform our competition." The Sause family has made a tradition of doing a good job in a quiet way for their customers, he said. "We think we do the job best." But they let their work speak for them and do not seek, in fact avoid, publicity.

Attempting to keep our conversation going, I mentioned the success of the Orville hook in recovering the *Nestucca* and said I'd heard it was now being made available to other towing firms. Sause explained that it is both sold and licensed to other firms for manufacture under the Sause patent.

Sause did contribute an intriguing bit of defensive information relative to the *Nestucca* spill. He said his experts informed him that no more than 2,000 gallons of oil reached the beaches of Canada. I asked how many gallons of oil he believes were beached in Washington, and Sause repeated "2,000 gallons." He asserted that no more than about 2 percent of the material removed from beaches was actually oil from

Nestucca. Such information was news to me.

I had one final request, to which Sause agreed. He saw no problem with providing photographs of the *Ocean Service*. At that time I had none, but later discovered two in Ecology's *Nestucca* file. Photos of the tug arrived in 1994, showing it moored at the Coos Bay headquarters.

In the days following our interview, I mulled Sause's contention that less than 2,000 gallons of oil had beached in Washington, and a similar quantity in Canada. Back in Olympia, I inquired of Dr. Jerry Galt, the NOAA expert on how oil spills spread. Galt said that the Sause figure might be reasonable for Canada, but not for Washington State. "There was probably 2,000 gallons of oil on the beach in the Norwegian Memorial vicinity alone," said Galt. As to Sause's assertion that only 2 percent of the material removed from beaches was oil, Galt suggested that would be low, but that 10 percent would be high for the oil content of beach debris removed following such a spill.

These observations caused me to return to official reports on the spill. The Washington State Department of Ecology reported more than 584 tons of oiled waste (sand, gravel, oiled debris such as sticks and wet vegetation/seaweed, etc.) was buried at the LeMay landfill, Aberdeen. About 45,000 cubic yards of oiled logs were burned, with the notation that the logs were less than 10 percent oiled. In addition, about 90 cubic yards of oiled pompoms were added to hog fuel and burned at the ITT Rayonier Pulp mill, Port Angeles (13-1).

Starting from Galt's opinion that 2 percent oil content would be low and 10 percent would be high, I took half the difference, 4 percent, and added it to the 2, making 6 percent. Then, giving Sause the benefit of any doubt and also keeping in mind that beach-cleaning crews were told to remove "anything that didn't belong there" including styrofoam cups and such, I cut back to 5 percent.

Using that percentage, the 584 tons of oiled waste collected in Washington State would represent 58,400 pounds or 29.2 tons of oil. Even at 2 percent, the 584 tons would represent 11.68 tons of oil, leaving aside the 90 cubic yards of oiled pompoms and the oiled logs that were burned. Not knowing how to translate tons of oil into gallons, I called our fuel oil dealer and was told that a gallon of oil weighs 7.35 pounds. Using that figure, the 29.2 tons of oil computes to 7,945 gallons of actual oil on Washington beaches, including those of Olympic National Park.

As to the oil on Canadian beaches, arriving at an estimate is more difficult. In the first place, there is Dr. Galt's opinion that Vancouver Island received "only spatter" and nowhere was subject to what he terms heavy oiling. Then there are the official reports from the Canadian Coast Guard and Environment Canada. The environmental report includes a table detailing the re-oiling of the Pacific Rim National Park beaches between Tofino and Ucluelet that indicates class 4 moderate oiling on Florencia Beach January 20, class 4 on Long Beach January 16 and 17, class 4 on Wickaninnish Beach January 15-18, class 4 on Combers Beach January 12 and on South Beach January 4 and class 3-4 January 12. Other days on the various beaches varied from class 2 to class 3, with class 3 medium reports giving way to light conditions after January 29 on all six beaches (13-2).

Criteria for the classes and oil cover conditions were:

Class 1: No visible oil on surface of rock or beach, none in subsurface sediments nor on logs.

Class 2: Sheen (rainbow oil) on surface but no visible black oil, and thin (<2 cm.) cover of oil on logs or rocks.

Class 3: Sheen on surface and brown or black oil blobs 2 cm. or less with occasional 14-20 cm. diameter blobs.

Class 4: Thick oil cover on rocks or beach (> 2.6 cm.) with blobs 25 cm. or larger and logs with blobs or oil > 2.5 cm. thick.

Class 5: Continuous oil cover, not blobs, > 5 cm. thick.

Oil cover ratings were a system of three. *Light*: 5 to 10 percent of shoreline covered with sheen, brown or black blobs 25 cm. or larger, no more than 10 blobs per square meter.

Moderate: Brown or black oil blobs 25 cm. or larger, more than one blob per square meter, and more than 10 meters of shore contaminated.

Heavy: Brown or black blobs 25 cm. or larger over more than 10 meters of shoreline, but not continuous cover.

The Canadian Coast Guard report says "450 tonnes (metric tonne weighs 2,204.6 pounds) oil waste material was collected. The oiled debris was eventually taken to a holding area near Nanaimo and then burned in the Ladysmith incinerator." Elsewhere in the report, however, it says "approximately 7,000 bags filled with oil and oiled debris...airlifted...by helicopter to containers at Pachena Village...were later burned on fires burning heavily oiled logs" (13-3). That was in the area of Pacific Rim National Park south of Barkley Sound, reaching to Port Renfrew along Vancouver Island's West Coast Trail, popular with hikers. Numbers of logs burned are not completely reported, but 2,000 were torched in one area, 2,600 in another, and in the Nootka Island area both logs and oiled gravel were burned together.

Using only the 450 tonnes and converting them to 496 U.S. tons, without considering the oil burned with logs, one gets a 5 percent figure of 24.8 tons. That in turn figures to 6,748 gallons of oil on Canadian beaches. But try Dr. Galt's opinion that 2 percent oil might be about right for debris on Canadian beaches, and the product comes to 2,699 gallons.

However, others related to the *Nestucca* spill also tried to figure how much oil was recovered from Washington beaches. Diane Harvester, Ecology beach supervisor, recorded in her notebook January 31, 1989, some best-guess calculations arrived at by Thom Davis of Global Diving & Salvage and John Weichert of CleanSound. Their total, 10,576 gallons, was higher than the 7,945-gallon figure arrived at above. Weichert and Davis employed a slightly higher weight per gallon, 7.5 pounds rather than 7.35, (bunker fuel is heavier than furnace oil) and also used a total of 252,000 gallons as the maximum spillage from *Nestucca*, but allowed for 20 percent evaporation. They estimated 1,000 gallons of oil recovered from Ocean Shores beaches, 376 from Sand Island, 450 at LaPush, 250 at Ozette, 500 at the Whale Creek dump and 8,000 gallons at Taffy Point using pompoms. That last figure was based on recovery of about 10,000 pompoms with an average oiled weight running from 5 to 10 pounds, but some as heavy as 25 pounds. Figuring in reverse reveals that they employed an average of six pounds of oil for each pompom. The mousse collected by pompoms is a mix of oil and seawater.

Taking the combined estimates for Washington and British Columbia beaches, one could say that recovered *Nestucca* oil may have been somewhere from a low of 5,877 gallons to a possible maximum of 17,324 gallons. That's excluding quantities of logs and other oiled debris burned. The original spill was between 223,104 and 231,504 gallons, so recovered oil would fall between 0.025 and 0.077 percent of the total spill.

Compared to the black floods that have plagued European beaches from some monstrous Atlantic oil spills, the quantities on Pacific Northwest beaches from *Nestucca* were small. But, as I remarked to Dale Sause, causing him to furrow his brow, "We sometimes learn quite a lot from small things."

For example, take Sause Brothers Ocean Towing. The company has purchased steel shipping containers at $45,000 apiece for each of the 10 barges used to haul oil or other bulk liquid cargo, plus a spare. Each container is equipped with 1,200 feet of oil boom, manufactured to 7-foot lengths rather than the standard 8 feet so as to leave walking space along one side of the van/container to get at all the items of equipment stored therein. Additional equipment includes a boat with outboard motor, 40 or more boxes of pompoms for picking up floating oil, large quantities of sorbents, plus rope, flashlights, hard hats, protective clothing, damage control material for patching holes such as the slash in the side of *Nestucca*, two-part epoxy, plywood, oakum, neoprene, plugs, a hacksaw, tape measures, thousands of plastic bags, tote pans, pitchforks, brooms, shovels, pads, PVC pipe, 4-by-4 and 2-by-4 lumber and a peristaltic pump.

Just purchasing the equipment and bolting the loaded containers onto company barges was not the end but rather just a beginning. The company now conducts periodic day-long training sessions for its employees, from President Sause down, to make certain everyone knows how to use the equipment. It was this program I had hoped Sause would allow to be publicized in more detail, because it represents why the veteran towing company is now considered a leader in its field. Even in Coos Bay, few people are aware how the local firm has taken responsible leadership in oil spill prevention.

One other new development reported relative to Sause Brothers Ocean Towing: the company has built a new oil barge, its first with a double hull, to succeed the *Nestucca*.

The *Nestucca* spilled no more than 231,000 gallons of bunker off Grays Harbor through its six-foot gash. It happened, ruled Judge Reddin of U.S. District Court, because of a defective tow wire and negligent maintenance and inspection. Except for the Orville hook, there could have been a 3-million-gallon spill, 13 times as large, had the barge beached and broken apart. *Nestucca* events cost the company and its insurors more than $15 million. Sause has since invested at least another million dollars in spill equipment and training for the future, outlays that could save far more than *Nestucca* cost.

But Dale Sause, in the family tradition, does not care to have the world know about company improvements, just their customers/clients. He runs a good operation, a company that is a responsible community citizen--and the Sause people intend to continue doing what they do well, and I can only hope they succeed.

* * *

Other companies may not emphasize advance planning as does Sause. Perhaps they have bad luck. It might be we pay more attention now out of awareness of the pollution threat lurking just over our horizon. It may be even closer, like down the road or out in the channel of a nearby waterway. But most often it seems to happen somewhere else, not at our doorstep.

Such incidents quickly pass in and out of public awareness, but the saga of the bulk freighter *New Carissa* prolonged itself indefinately after it grounded on the ocean beach while awaiting entry to Coos Bay, Oregon, Feb. 4, 1999. Coincidentally, that was but a couple miles from Sause headquarters. (Further details are in chapter 15.)

Perhaps even more exceptional, but fortunately less tragic for the vessel, was the grounding of the Russian cargo ship *Monchegorsk* (13-4). It ran ashore in front of Anderson Island residences northbound from Olympia toward Tacoma the evening of September 5, 1998, after the Puget Sound pilot inexplicably turned off course, according to a Coast Guard report. That is some two miles south of where the ship would normally have rounded the island to head past Tacoma en route back to Russia. Heavily reinforced for ice breaking, the *Monchegorsk* was undamaged as it slid onto the mud shallows just offshore.

"Island residents, basking in the sun on their decks around 7 p.m., were stunned" as the huge vessel surged toward them, according to *Tacoma News Tribune* reporter Al Gibbs. The ship was pulled free on the high tide next morning and inspected in Tacoma before being allowed to continue her voyage. The pilot was reprimanded following a hearing before the Puget Sound pilotage commission and fined $3,000, half held in abeyance pending remedial training. He paid the $1,500 and then chose to resign.

But if such events seem unusual or abnormal, the folks at Washington State Ecology have tallied other near-accidents (13-5). During the winter season of 1998-99, on Nov. 24 the tank barge *Columbia* went adrift when its tow cable parted in heavy seas off Cape Shoalwater, at the entrance to Willapa Bay (between the Columbia River and Grays Harbor). The tug was able to re-establish the tow and proceed to Port Angeles.

The very next day, Nov. 25, the cargo ship *Aristotelis* lost propulsion three miles west of Cape Flattery and drifted across the strait toward Vancouver Island, where it was able to anchor, preventing another grounding. Assistance from a tug was not available.

Jan. 18, 1999, the bulk carrier *Sabina* (formerly *Irenes Vision*), lost propulsion off Stuart Island, in the San Juans proximate to the border of British Columbia, and drifted for two hours in Haro Strait.

February 5, the container ship *Apljapan* lost propulsion off Port Angeles and drifted within 400 yards of the beach at Ediz Hook before the crew was able to restart the engine as a tug stood by.

Then March 28, the manned tank barge *Bouchard 230* and tug *Ralph Bouchard* ran into heavy seas off Cape Johnson on the Olympic Peninsula, and its tow wire parted. After drifting some 8 miles, the barge was met by a rescue tug, which stood by while the *Ralph Bouchard* reconnected to the barge about 11 miles west of Point of Arches.

Mariann Cook Andrews, editor of *Spill Scene*, asked rhetorically, "Are we pushing our luck?"

14

A New Day for Oil Spills

The 1990 *Nestucca* pre-trial court settlement between Sause Brothers Ocean Towing and the State of Washington and U.S. federal government provided that Sause, as the responsible party, would pay $3.35 million to state and federal agencies: an initial $1.35 million for response costs during the incident, plus 10 annual payments of $200,000 totaling $2 million.

Results from employing those dollars have improved the situation in Washington State to national leadership status in oil spill prevention, preparedness and cooperative planning.

Curt Smitch, director of the Washington State Department of Wildlife (now the Department of Fish and Wildlife), said at the time, "The settlement is very fair and gets the state moving in the right direction. Creation of the response fund is particularly fitting. (It) will be a legacy of the cooperation between the state and Sause...in reaching a fair settlement that also provides a program to lessen the impact of any future spills along our shores."

Contingent valuation methods developed during the negotiations were also cited by Smitch. State and federal wildlife biologists employed the new tool in evaluating wildlife damages caused by the spill. Smitch termed "CVM" the leading edge in resource damage assessment. At the same time, he praised the Coos Bay, Oregon, towing company for its quick and responsible actions during and following the accident, saying, "Sause Brothers has been a great corporate citizen since the minute the accident occurred. Their commitment to taking care of our environment is unquestioned."

Dale Sause, company president, also was quoted at the time. "Since my family began Sause Brothers three generations ago, we have always worked hard to do the right thing. The 1988 oil spill was a very unfortunate accident, but we worked hard to make it right. We hope that the emergency response fund will leave something positive in its wake."

It has. Not everything positive that has grown out of *Nestucca* can be called a result of the financial settlement. Of course, it is notable that there can even be good results from an event as negative as an oil spill. One positive aspect of the event has been the learning in various quarters that has resulted from the experience. The pre-existing spill response team that had been created in the Washington State Department of Ecology gained invaluable experience, and cooperative unified supervision of ensuing spills through the Incident Command System is another spill management gain.

Major improvements have been the state's interdepartmental Natural Resource Damage Assessment Team (NRDA), advance planning for resource protection and preparations for wildlife rescue and response, financed by *Nestucca* funds. NRDA Pre-assessment Screening And Oil Spill Compensation Schedule Regulations have become a national model (14-1).

This has proven to be a useful tool in Washington State, a source of income for an unglamorous but vital program that otherwise would have little claim on the state general fund. Contingent Valuation Management (CVM) has been mentioned as an innovation. The NRDA program is more creative, and deserving explanation despite its technical nature. We do so in chapter 15.

International cooperation that has come about as a result of *Nestucca* can be credited in considerable degree to Canadian citizens' outrage over invading oil, slow official reaction and response, and finally to the willingness of political leadership. Commercial consciences wounded by oil spills also contribute financial support, not without self-interest in seeking to reduce the likelihood of multi-million-dollar fines and court judgments resulting from future oil spills. Evidence of this is the support for improving CleanSound Cooperative, Inc., and also the newer national cooperative, Marine Spill Response Corporation (MSRC), both of them oil recovery organizations with specialized vessels and crews. Sibling groups have been or are being developed up and down the Pacific Coast and elsewhere.

Perhaps the biggest single aspect of the entire picture that has developed since *Nestucca* and *Exxon Valdez* can be summed up in a single word: Cooperation.

Major oil spills are of a size and scope that no one entity, such as the U.S. Coast Guard, can cope with the ramifications. Management of even a medium-size spill such as *Nestucca* involved, in addition to the Coast Guard, the U.S. Fish & Wildlife Service, the National Park Service, Washington State Department of Ecology, State Wildlife and Fisheries (now consolidated), State Department of Natural Resources, plus Emergency Management and the Washington Conservation Corps; Grays Harbor, Jefferson and Clallam Counties; the Quinault, Quillayute and Makah tribal governments and the City of Ocean Shores. That's for starters, plus Sause (SBOT), International Bird Rescue, Global Diving and Salvage and another beach cleanup and salvage company, Chempro, now part of Foss Tug & Barge subsidiary

Foss Environmental, Eagle Air helicopters and others. Not only these, but their counterpart government ministries and private contractors were involved in British Columbia, and one can't exclude the thousands of volunteer workers and tribal people involved in both Washington and Canada.

Cooperation was a key word at the time, in late 1988 and early 1989, and continues to be. But it must go hand in hand with the financial wherewithal for planning, and providing training and equipment to carry out those plans. That brings us back, then, to the legacy of *Nestucca* to which Smitch referred.

The initial settlement payment of $1.35 million included $320,000 for the United States Department of the Interior to repay response costs of *Nestucca* incurred by the Coast Guard, the National Oceanic and Atmospheric Administration (NOAA) and also for claims by the Hoh, Quilleute, Makah and Quinault tribes, through the Bureau of Indian Affairs. The balance of federal payments, $500,000, is being made in 10 annual installments of $50,000 starting in 1991.

U.S. Fish & Wildlife Service environmental contaminants specialist Jeffrey J. Momot of the Olympia district office wrote the April, 1995 study and response plan for employing the funds that had been accumulating for restoring the damaged, destroyed or lost natural resources under trusteeship of the Interior Department (14-2). The plan includes public education, habitat restoration and alternative measures to reduce seabird mortality, and monitoring common murre survival and reproductive success.

The most specific action selected was eradication of rabbits from Destruction island, off the Olympic National Park coast between Queets and La Push. European rabbits had been introduced during the early 1970s, destroying or occupying the breeding habitat of burrow-nesting seabirds such as the tufted puffin and rhinoceros auklet, and also cropping vegetation and accelerating erosion of the island.

Education is aimed to increase public knowledge of boaters, fishermen, pilots of planes or helicopters and all visitors to national wildlife refuges of the need to minimize disturbance to seabirds and other wildlife. Educational brochures, posters and interpretive signs will be augmented by large warning signs on both Destruction and Ozette Islands advising boaters to keep a minimum distance of 200 meters (600 yards) from shore.

Net fishery alternatives include further investigation of utilizing visible mesh in the upper part of gillnets to reduce seabird mortality and seeking other options in cooperation with government agencies and tribal fisheries.

Monitoring will increase knowledge of various factors affecting murre population. Time lapse photography and record keeping on breeding colony disturbances will be included along with regular patrols and enforcement action as needed.

Washington State's initial $1,030,000 was divided, with $876,000 to the Wildlife Department and $154,000 to the Ecology Department. Of the Wildlife portion, $516,000 was for out-of-pocket recovery costs, and $360,000 was set aside for a permanent rehabilitation center and for primary care units and trucks or trailers to carry oiled birds to the center. That was further divided, and $250,000 retained for establishing the permanent bird center, leaving $110,000 which is being spent for rescue units and transport trailers. One van has been equipped as a transport unit and a 40-foot highway trailer constructed for use as a primary-care center for oiled birds at the site of a marine oil spill. It now is assigned to Neah Bay (see chapter 15).

The ten annual payments totaling $1.5 million provided by the settlement to the state of Washington include $74,000 for wildlife rehabilitation, $50,000 for Washington coast research and $26,000 a year for Ecology's response capability.

It is the programs these dollars made possible that have moved Washington State to the forefront in preparing for the inevitable future oil spills. There are also two other sources of funds, the $0.05-per-barrel oil spill response tax (14-3) enacted in 1991 for collection from companies handling oil, and the dollars that come in from NRDA payments by those responsible for smaller spills.

There is more. The Washington State Office of Marine Safety (OMS) has been created, and during 1993 began screening cargo and passenger vessels that enter state waters to determine which may pose a substantial risk to health, safety and the environment. New department rules require that all tankers and some other vessels submit advance notice of entry, in addition to that provided to the U.S. Coast Guard, and to report unusual events. Four regional marine safety committees established under OMS have submitted safety plans and recommendations. These cover such things as vessel traffic, environmentally sensitive areas, seasonally risky situations, requirements for tankers and perhaps other vessels to have tug escorts, and a need for more and better communication by and between vessels. Non-English-speaking mariners, for instance, were noted as a problem requiring attention.

New bunkering rules have been adopted, and also standards for best achievable protection and technology for tank vessel operating procedures, personnel and management practices (14-4).

In 1997, pursuant to legislative action, the Office of Marine Safety was merged with the Department of Ecology, combining the marine vessel safety and oil spill prevention activities with Ecolgy's prevention and statewide oil and hazardous material response and restoration program. Joe Stohr was named to lead the new Spill Prevention, Preparedness and Response Program at Ecology, where he previously had worked in water resources and on nuclear waste.

Ten years after the *Exxon Valdez* grounded and spilled 11 million gallons of Alaskan crude oil in Prince William Sound, two men with first-hand knowledge who now work in Washington's spill program agreed that this state's level of spill prevention, preparedness and response is among the best (14-5). Paul O'Brien, who heads spill response at the Northwest Regional office, cited the Geographic Response Plans, first in the country, as a "priceless tool...(that) puts us way ahead of the curve in knowing what to do first. In a response situation, timing is everything."

Roy Robinson, contingency plan reviewer at Spills Program headquarters, agrees, adding that Washington's preparedness program has placed equipment where it can be available for rapid response. Both agree that the program of drills is key to the state's high level of capability, helping insure that both companies and government agancies are better prepared to respond quickly and effectively. "We've built a good, strong working relationship with the regulated community," said O'Brien. Lack of the incident command system "was a big flaw in the *Valdez* respone," Robinson noted. "Don't get complacent," he continued. "The lesson we learned (in Alaska) is prevention is a lot better than response. Our emphasis should be placed on preventing accidents."

Vessels visiting Puget Sound, whether foreign or domestic, are required to prove their ability to provide effective response to possible incidents while discharging or taking on either bunker fuel or petroleum or other hazardous cargo. CleanSound, the industry-financed oil spill response cooperative, handles a lot of that responsibility through contracts with shipping companies.

CleanSound Cooperative, Inc., with headquarters at Edmonds, provides part of the oil spill response matrix for Washington. Executive Roland Miller says its genesis was the Santa Barbara, California, oil well spill events of the early 1970s. "From that spill, it became evident to industry that there was a need for group organization external of themselves," recalls Miller.

"The first actual cooperative...was Clean Bay Cooperative in San

Francisco, and...from that they spread...through Western Oil & Gas Association (WOAGA). It was a West Coast formulation that now is spread throughout the United States and Canada. CleanSound was established in 1971, and was called Washington Oil Spill Cooperative Association. It didn't change its name until 1973. There were 10 or 11 members in the original group, and...first they hired a contractor,...the old MOPS group," Miller continued.

"Around 1975...the co-op...(started) having its own resources under the ownership of CleanSound,...(and) started building... equipment. In 1990 we became a corporate structure...because it has to work today like a business. Because of major growth...and some foreign ownership in CleanSound, we...provide some... protection to members, and...other legal ramifications. Along with that, when you take a CleanSound with 30-some vessels, just the maintenance--you have a full time maintenance person and workers--you have personnel issues. You're not just a simple little organization, and they're not just little bitty boats. The biggest one I have is a 26,000-barrel barge (1,092,000 gallons), and I have a 126-foot ocean-going skimmer vessel. They all have to meet Coast Guard requirements and regulations, so we schedule inspections and such.

"The annual budget is close to $3.5 million (including) maintenance, salaries, insurance, everything. It's about a $25-$27 million organization (capitalization value). We depreciate equipment on a regular schedule, and have a replacement fund. Financing is by a system of taxing the business members on the basis of the number of barrels of fuel they process or handle. Some members are in the 20 percent (taxation/contribution range) and some (at) half of one percent. You take your operating expenses, divide that number in and charge them their share. We don't audit them, they have to be honest, report to us their statistical records," Miller explained.

Facilities members of CleanSound include ARCO Products, BP Oil, Chevron USA, Olympic Pipe Line, Shell Oil, Texaco, Time Oil, Trans Mountain Oil Pipe Line, UNOCAL and U.S. Oil & Refining. Marine members are ARCO Marine, BP Oil Shipping, Chevron Canada and Chevron Shipping USA, Foss Maritime, Imperial Oil, Sause Brothers Ocean Towing, SeaRiver Maritime, Shell Oil Marine Dept., Texaco Marine Services and West Coast Shipping.

"A lot of people think CleanSound membership is (only) here," observed Miller. "But I have members in Houston, Long Beach and other areas. They process oil through here. And now we're dealing with a lot of foreign entities, because of...new laws requiring them to have...a contractor in place (in case of a spill). So we're dealing with

Kuwait, Holland, Taiwan, Hong Kong and other places worldwide where companies send ships into Puget Sound.

"From here to Alaska, and (south) to the California border, you have two voids (in coastal protection)," Miller said. "From the Columbia River south there is no major resource in terms of a co-op (except for a local Coos Bay group). And from the tip of Vancouver Island (Cape Scott) north to Ketchikan there is no major resource capability. In the Alaskan panhandle you have the Southeastern Alaska Petroleum Association (SEAPRO). Not a lot of resource base yet, but they're there. Now supposedly Marine Spill Response Corporation (MRSC), our counterpart, will...look at the Oregon coastal area. They're going to put a program into Astoria. There is some equipment in the Columbia River, but that is not (capable of) offshore (operations).

"Up and down the coast there are four major co-ops (similar to CleanSound). Burrard Clean to the north (in British Columbia), then to the south the next is Clean Bay (San Francisco), then Clean Seas in Santa Barbara, then Clean Coastal Waters in Long Beach. Up north you have Cook Inlet Oil Spill Cooperative and Alyeska, a real big one. (The rest) are in the same range we are.

"I think what we're learning is that we must clean up our act," asserted the CleanSound executive. That sort of statement was unlikely until recent years from officials related to the oil business.

Miller mentioned some oil spills not previously alluded to in this manuscript as he commented on the *Nestucca*. "It was a major spill for the coastal area. Before *Nestucca* we had the *Blue Magpie*...off Newport, Oregon. A...Japanese cargo ship...missed the (harbor) entrance and went onto the north side of the jetty, (spilling)...80,000 gallons of oil, and some of it went north. Fortunately it didn't hook around and come back onto the beach. It was in November of 1983.

"The only other...that's impacted the beaches...was the tanker *Mobil Oil* when she hit (a) rock on the Columbia River. That was about 200,000 gallons, and quite a bit of that came out of the river and washed up...on Long Beach..., mostly little tarballs, not a major thing. I think what made *Nestucca* more unique...was...that it impacted the national park and Canada. The most recent was the *Tenyo Maru* in 1991,...where the oil went south in summer. There have been some other spills. You have Texaco pipeline at the refinery (Anacortes, February, 1991, 210,000 gallons), which we worked on, and...U.S. Oil in Tacoma (600,000 gallons, January, 1991), but those were on land."

Commenting on the regular practice drills CleanSound takes part in, Miller said they involve crossover work with Canadian personnel because of the proximity. Queried whether Canadian authorities are better prepared than in 1989, he said, "Well, it's interesting, they're going through an evolution. They're not up to where we are. But...the Canadian Coast Guard (formerly) was always up front and had ...responsibility for taking the lead. Now (they have)...dropped back to a supervisory stance, and they're going to force industry to take a more proactive role. Our counterpart, Burrard Clean, has quite a bit of equipment, but...only a few people. This change may force them to increase. Plus, they're only talking about Victoria and Vancouver now, but how about Prince Rupert and the other outlying places?"

CleanSound maintains vessels and full time crew members at seven ports around Puget Sound, including the seagoing *Shearwater* at Port Angeles and the *Arctic Tern* at Port Ludlow, both on Juan de Fuca Strait, and other vessels of varying size and capability at Bellingham, Anacortes, Edmonds, Seattle and Tacoma. Crewmen are supported by trained Industry Response Team members from all five oil refineries in the area, plus more than 50 contractor crew members on call.

The industry-financed corporation declares it is prepared and cooperates fully with federal, state and local government agencies to assure that in any spill the cleanup priorities are met while preventing environmental damage as much as possible.

Another aspect of oil spills I've not discussed involves those that occur on land, and new state regulations cover advance planning and training for personnel at all oil-handling facilities as well as those involved in maritime transport.

Oil spill preparedness, however, is more than just statewide. Federal and state agencies have a contingency plan covering Washington and Oregon marine waters, but also extending to Idaho. Participants include the U.S. Coast Guard's Portland and Puget Sound Marine Safety Offices, the U.S. Environmental Protection Agency, Oregon's Department of Environmental Quality, and the Washington State Department of Ecology. This plan reinforces use of the Unified (or Incident) Command System as a consistent way to organize varied agencies with differing responsibilities into one concerted response effort in case of a spill. It also envisions use of geographic response plans, site-specific information prepared in advance for any spill occurrence.

The next level of planning/preparation/cooperation involves the States/BC Oil Spill Task Force, a group composed of British Columbia, Washington, Alaska, Oregon and California. It was organized following the *Nestucca* spill by British Columbia Premier William Vander Zalm and Governor Booth Gardner of Washington in response to frustration over the oil spreading into Canadian waters, then enlarged to include the other Pacific Coast states. But the evening following the first task force meeting in 1989, the tanker *Exxon Valdez* ripped the bottom out of her hull-full of Alaskan crude oil on Bligh Reef in Prince William Sound, Alaska. Some 11.1 million gallons of crude escaped, and the birth of the new international group went unnoticed.

The States/British Columbia Oil Spill Task Force has taken on a life of its own. "Initially," said Ron Holcomb, Washington's Ecology Dept. spokesperson, "it was just officials from Ecology and our counterparts from the British Columbia Ministry of the Environment, and our focus was just on our two areas. The first thing the task force really participated in was when Exxon wanted to bring the *Exxon Valdez* down to Portland for repair. That brought in Alaska and then Oregon because of the problems involved in transiting a damaged tanker. We spent hours negotiating with Exxon and the Coast Guard on what conditions would be acceptable for transit up the Columbia River. We agonized over would it even be appropriate and even hired an expert from Scandinavia, one of the world's premier marine engineers, and sent him to Valdez to determine what was being done to stabilize the tanker for the voyage. And after we worked out the conditions, Exxon realized it would be easier to take it down to California.

"Since then, with the addition of California," continued Holcomb, "we've been working together as a unified West Coast group focusing on marine shipments. As the interim report of December 1989 and the plan of August 1991 were finalized, there were joint recommendations for all of the states and British Columbia and in addition individual recommendations for each state and the province."

As the task force's action plan was announced in 1991, the five departmental directors asserted in a joint statement, "We ...now...embark on another major step forward (with)...focus on cooperation between the member governments to ensure consistency in our...marine oil spill prevention and response policies and regulations, in order to minimize the burdens placed on industry without compromising the scope and effectiveness of the standards.

"In view of the *Exxon Valdez* disaster and other major oil spills in recent years, we and our mutual federal governments must never again become complacent about the potential for environmental catastrophe," continued the statement. "Prevention measures provide by far the most effective means of protecting our valuable waters, shorelines and natural resources, and the contributions they make to our economies."

The first major milestone of the task force was compilation and publication of its 1990 "Final Report" with six major findings and 46 recommendations (14-5). The major findings included significant problems revealed by recent spills: inadequate personnel training and qualifications, shortcomings in vessel design and integrity, insufficient traffic management, gaps in regulatory oversight and incomplete cost recovery by states/provinces.

Other significant conclusions were recognition that little oil can be recovered from a catastrophic spill, and therefore preventing spills must be the prime strategy. Readiness and response to smaller spills must still be emphasized. Further, comprehensive oil spill prevention demands wide participation, and the task force will continue to promote coordination of West Coast oil spill prevention and response efforts.

Priority groupings of the recommendations resulted in a primary list headed by petroleum conservation, followed by alternative transportation, vessel traffic service systems and vessel safety measures. Then followed double hulls, onboard navigation improvements, petroleum facility worker training, mariner qual-ifications, crew requirements, strong sanctions and proof of financial responsibility, plus others.

The second priority list was led by the need for tug crew training, natural resource valuation, cost recovery, response plans, clean-up requirements, response training, wildlife rescue training and equipment, response drills and others.

A third priority list included tug escorts for single propulsion and very large tankers, a near-miss report system and adopting requirements for towing cables and towing systems, plus additional recommendations. A new rule requiring tugs to escort large oil tankers through Washington's inland waters became effective in November of 1994.

Relative to double hulls for oil tankers, the task force report's rationale takes note of a 1975 U.S. Coast Guard study that concluded double bottoms would have contained some 90 percent of the spilled oil (14-6). Another study in 1990 estimated 36-50 percent reduction in spill probability with double hulls (14-7).

Requiring all tankers to operate on-board navigation equipment such as Electronic Chart Display Information System (ECDIS) might provide a 20-30 percent reduction in oil spill risks (op.cit. 14-7).

Perhaps most important among the recommendations is for more stringent mariner qualifications, based on the rationale that "up to 90 percent of all tanker casualties occur due to flaws in human performance in either operation or maintenance" (14-8).

Representatives from the five governments making up the task force adopted short and long term objectives and goals, and opened an office in Portland, Oregon, during 1993, with Jean Cameron as executive coordinator (14-9).

15

Washington Sets Example

Washington State's preparedness, operational and followup programs for coping with oil spills both large and small deserve attention in some detail. This involves mainly the departments of Ecology and Fish & Wildlife, but also Emergency Management (15-1), and on the county level the sheriff or his emergency management officer. Of course, the Coast Guard is the federal agency to be notified in case of any oil spill in marine waters.

To report a spill anywhere in Washington State, the 24-hour telephone number is 1-800-258-5990. A responsible party should notify the duty officer at that number (15-2). Ecology then will be notified to begin initial response action from one of its regional offices providing 24-hour statewide coverage. The Seattle Coast Guard number is 1-206-217-6232, although marine spills south of Queets come under the authority of the Portland Coast Guard District, phone 1-503-240-9379 or 9301. Ecology and the Coast Guard work together as the lead state and federal agencies to oversee the responsible party's cleanup efforts.

But before an oil spill happens, a lot of preparation has been put in place. The entire coastline of Washington State, including the Columbia River, has been segmented, and each piece of shoreline reviewed by the Natural Resource Damage Assessment Team. The team is composed of representatives from all state resource agencies, and from the Department of Health and the Office of Archaeology and Historic Preservation. Both state and federal agencies, along with county and local (city or town) and tribal governments, have worked along with the Coast Guard, oil industry people and spill response contractors in planning sessions to identify sensitive areas and develop protection strategies based on local conditions at any given state of the tide or time of year.

Geographic Response Planning for each segment of coastline deals with three main concerns: natural resource protection, response strategies and logistical support.

Dick Logan of Ecology, chair of the NRDA team, provided a sample copy of the Washington Geographic Response Plan (GRP) covering the Strait of Juan de Fuca and Cape Flattery (15-3). It's fascinating in its presentation of specific information keyed to local area map segments, with oil response strategies in detail for each piece of coastline.

Let's look at a couple of computer-generated examples of map segments (see figs. 7 & 8, pages 251 & 252), and the key explanations which each official responding to a spill will have in hand, whether that person be from the state, a county, the Coast Guard, a contractor working under the responsible party (who pays the bill for cleanup) or perhaps a tribal entity.

The first map (fig.7) shows Cape Flattery and Tatoosh Island offshore. Notice first that almost the entire area is included in the Makah Indian Reservation, and keep in mind that access and all management decisions relative to a cleanup must have the approval of the responsible tribal official (15-4). Next, you will see in Neah Bay the mixed sand and cobble beach, the exposed rocky shore (much of that shoreline high cliffs), extending around Cape Flattery and also Tatoosh, continuing to the mouth of the Waatch River. Its marshy shoreline changes to gravel/cobble as the coast continues southward beyond the reservation and into Olympic National Park.

Next are those circled numbers, some of them with arrows. These refer to a variety of protection/collection options, which are listed on separate pages of the notebook keyed to the map. S-1 lists "dispersants, case by case basis, fall season only, to break up spill, (with the) objective strategy (to) prevent oil from reaching shore. Intent of (the) strategy is to disperse oil slick with chemical compounds." In a final column reserved for comments is this: "Permission to use dispersants must be obtained. Dispersants force oil into water column, which may be harmful to offshore fisheries."

S-2 refers to *in situ* burning, again on a case-by-case basis, to prevent oil from reaching shore, with the comment "Permission must be gained from local air quality authorities. Air pollution is a likely outcome of this strategy." S-3 is open water skimming along the shoreline, again to prevent oil from reaching shore. This strategy, implemented at the discretion of the on-scene commanders, would employ five 1,000-foot sections of collection boom with one to five coastal skimmer vessels.

Off the mouth of the Waatch River is the notation S-4, an exclusion strategy to prevent oil from entering the river mouth. It would employ a 500-foot boom blocking the river to protect salmon upstream. The strategy notes that access is by road from Neah Bay.

The S-5 strategy for Tatoosh island specifies *in situ* burning or dispersants if possible, otherwise use pompoms to prevent oil from reaching shore. Under "intent of strategy" is printed "Adsorb oil before it reaches shore. Use 1,000's of feet of pompoms as needed." Apparently the pompoms are to be strung on lines as they were

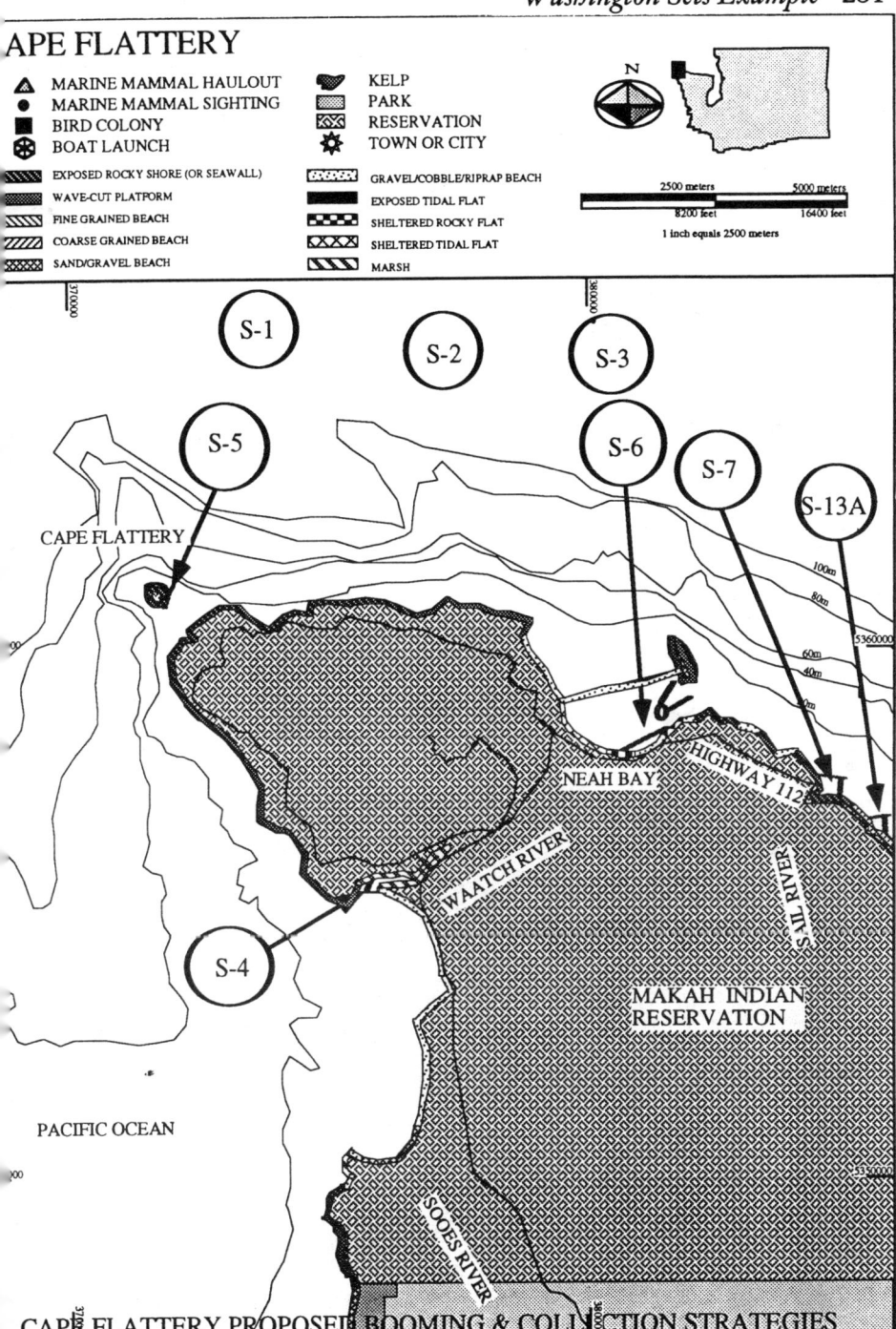

APE FLATTERY

△	MARINE MAMMAL HAULOUT		KELP
●	MARINE MAMMAL SIGHTING		PARK
■	BIRD COLONY		RESERVATION
⊕	BOAT LAUNCH	✿	TOWN OR CITY

EXPOSED ROCKY SHORE (OR SEAWALL) GRAVEL/COBBLE/RIPRAP BEACH

WAVE-CUT PLATFORM EXPOSED TIDAL FLAT

FINE GRAINED BEACH SHELTERED ROCKY FLAT

COARSE GRAINED BEACH SHELTERED TIDAL FLAT

SAND/GRAVEL BEACH MARSH

2500 meters   5000 meters
8200 feet   16400 feet
1 inch equals 2500 meters

S-1 S-2 S-3

S-5 S-6 S-7 S-13A

CAPE FLATTERY

100m
80m
60m
40m

NEAH BAY HIGHWAY 112

WAATCH RIVER

SAIL RIVER

S-4

MAKAH INDIAN RESERVATION

PACIFIC OCEAN

SOOES RIVER

CAPE FLATTERY PROPOSED BOOMING & COLLECTION STRATEGIES

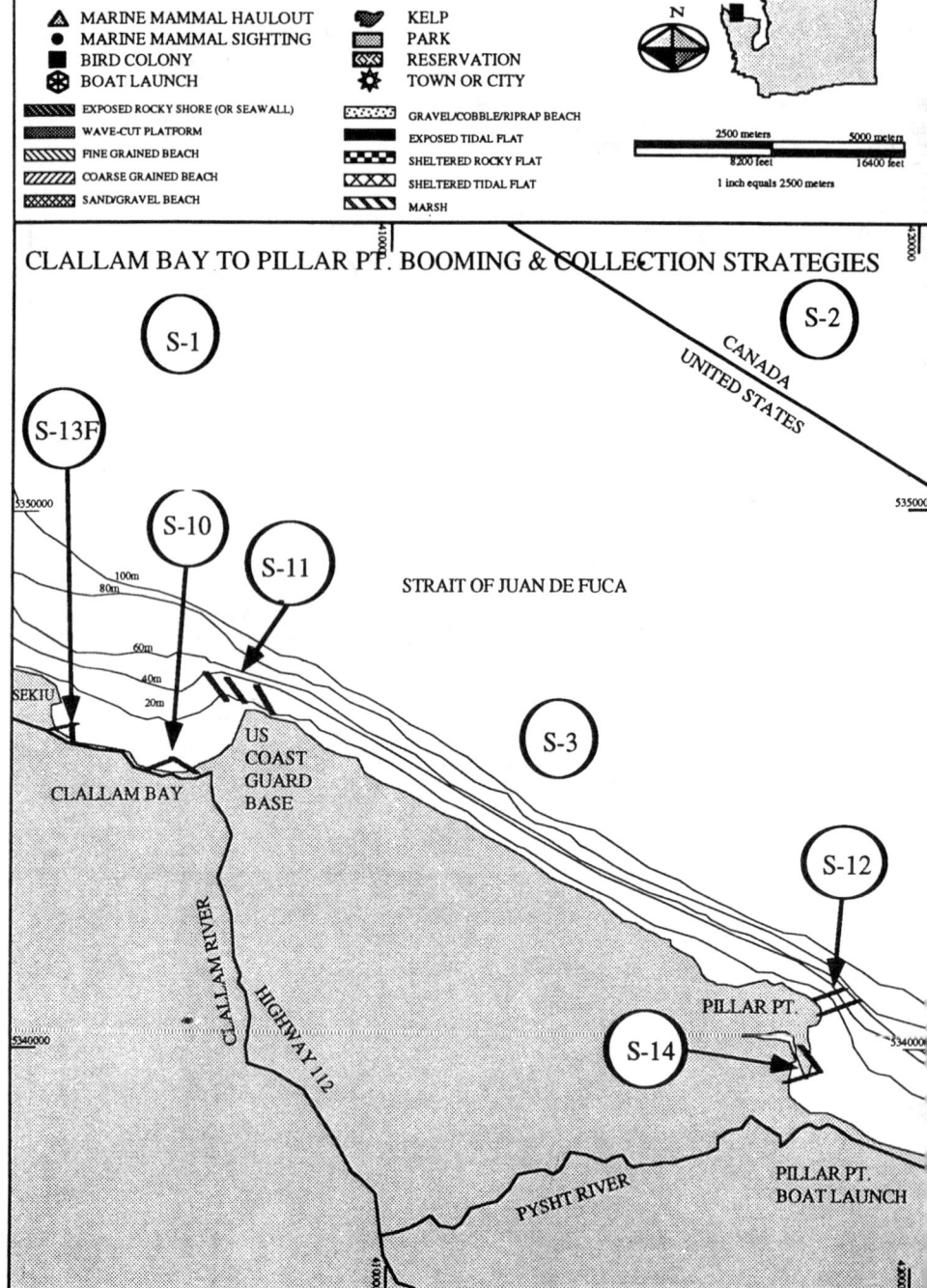

CLALLAM BAY TO PILLAR PT.

employed during *Nestucca* along the Olympic National Park shoreline south of here. Under the column listing resources requiring protection are common murres, Steller sea lion, harbor seal and storm petrels. (Recall that the "common" murre is no longer common in this area.) Under the comment column is this: "Site has high water action, strong currents. Rocky, inaccessible cliffs and shoreline."

S-6 for Neah Bay has a "Priority 2" listing, while the previous places have all carried "Priority 1." Response strategy is listed as deflection/collection with the objective to protect archeological areas, utilizing 3,000 feet of exclusion boom from the USCG station to the beach and 2,000 feet of boom for skimming if oil is on outgoing tide. That little circle with the V in the harbor indicates a skimmer. Steller's sea lions, northern sea lions and shellfish are resources noted for protection. A resource person to identify protection areas is required, and in Neah Bay there is a staging area listed with the added note "boat cleaning area, skiffs, wipe down boats." Under comments is "Site is archaeological sensitive--cultural, fishing, tourism. Bay is a natural collection area. Do not close harbor-staging area! Boom (is) available at Neah Bay."

There are two other designations, S-7 for an exclusion boom at the mouth of Sail River and S-13a, for the mouth of Snow Creek, with the notation "Coho (salmon) February/June out-migration."

One other component of the map is the offshore depth contours, which are given in meters (one meter is roughly equivalent to 39 inches). This is a rather steep-walled underwater canyon to the north, Juan de Fuca Strait.

For a somewhat different area, look at Clallam Bay, a few miles eastward (fig. 8). The first thing to notice is that the shoreline component overlay has been omitted on this map, so one can view the shoreline plain; but in fact the beach is sand and gravel in Clallam Bay and fine sand along the shore to Pillar Point, where it changes to tideflat and marsh. The jagged black line labeled Pysht River is a bit confusing, because it also represents Highway 112. The Pysht River estuary is as indicated by the S-14 arrow.

Again in this vicinity, S-1, S-2 and S-3 are available strategies, depending on conditions, as mentioned previously. S-13f notes chum salmon out-migration, February-June at Falls Creek. S-10 calls for an exclusion zone with a 500-foot boom to prevent oil from entering the Clallam River, affecting salmon. A comment notes the presence of a kelp bar, and says the river dissipates behind the bar in summer. S-11 shows 1,000-foot diversion booms deployed off the USCG base at Slip Point to protect shellfish in Clallam Bay. S-12 at Pillar Point shows more deflection boom to keep oil from the Pysht estuary, noting "very sensitive shellfish resources" and adding that shallow water skimmers can be used only during flood tide in the estuary behind the sheer cliffs of this archaeologically sensitive vicinity. On these shores no beach cleanup is allowed until authorized by the Washington Archaeology and Historic Preservation office (15-5).

This one notebook includes 10 similar maps with overlays, each with general and specific instructions designed to guide the first 12 to 24 hours of an oil spill response. Information includes command center locations, notification lists, location and availability of response equipment, trained personnel and facilities/services, hydrology, currents and tides, winds and climate information as needed, and the resource information and response strategies as sampled above. In addition to Juan de Fuca Strait, other GRP areas are the San Juan Islands, North Puget Sound, North Central Puget Sound, Central Puget Sound, South Puget Sound, Hood Canal and the Outer Coast. To these eight GRP areas are added GRPs for the Columbia River from its mouth to Tri Cities and the mouth of the Snake River, plans developed jointly with Oregon. In the future, plans will be prepared for the upper Columbia and tributaries such as the Spokane and Snake Rivers, and perhaps others, according to Logan. Some lakes and reservoirs might also be covered.

All these GRPs are available on computer for printout as necessary for crews responding to spills anywhere in the state, updated with the latest information. Periodic practice drills test response readiness for possible marine spills in different areas.

Now we turn to response preparation and training under the Wildlife Department, the program initiated using *Nestucca* funds. Key persons involved are Leni Oman, habitat management spill response coordinator (14-6), and Sara LaBorde, spill response rescue and education planner. "Sara is a lady who is just dynamite," I was told by Oberlander, the spill response veteran. Indeed, LaBorde is a living demonstration of enthusiastic expertise, and of course is backed by dozens of biologists and other expert functionaries of the Fish & Wildlife Department. Oman, the supervisor, has been on the job since May of 1992, LaBorde since July that same year. Oman has since tranferred to other responsibilities. Previous to her present job, LaBorde was liaison for the Wildlife Commission, and had developed the Wildlife agency's public involvement program working with the information division. She was later reassigned as special assistant to the director of Fish & Wildlife. Mick Cope heads the department's spill response program, and Chris Lane works with wildlife rehabilitation, and the training of workers and volunteers.

For the moment, review Oman's information concerning the program financing. As she joined Sara and me in 1993 with my tape recorder, I had asked whether the 5-cents-per-barrel state tax helps support the high cost of basic research, such as investigating the effects of oil on wildlife. Leni explained that the tax on oil transported in Washington harbors is split into two accounts. "There is the response account, which is used only during response to spills of magnitude; and the other side is the administrative account that funds 10 agencies or programs within agencies. It pays for routine spill responses, administrative costs to get the programs underway, coordination. There are seven components, (but) not a specific (category) for research. It requires us to be knowledgeable, and that seems to (require research). The other place that research comes in is...the coastal protection fund. When we have spill events, whatever size..., the state recovers damages for injury to the natural resource. Money goes into... the coastal protection fund,...regulated through (its) steering committee. Those funds can be spent on restoration for...injury at the site, and off-site studies can be done. There are four... components, one of (which)...is long-term impacts of oil on the environment. But for baseline inventory to (establish the prior background proof of) long-term injury,...there's a shortage of money right now," Oman said.

"When we talk about baseline inventories, it's identifying what the natural resources are, where they are,...(their) magnitude and...the stability of that population over time. It's a huge program. We do have some data on baseline resources. But we have a wonderful, beautiful coastline that...hasn't been researched very thoroughly,.."

"As to the *Nestucca* dollars," she continued, "...Wildlife received...a onetime payment...of $360,000 for wildlife rescue and rehabilitation equipment. We had some room to define what that meant, and put part into primary care trailers. It also includes $250,000 seed money (reserved) for the permanent rehabilitation center facility. St. Edward (state park on Lake Washington) is an interim use until we decide (whether to remain there or create another permanent facility).

"We receive annually (for 10 years) $124,000, (of which) $50,000 is for assessing the state...coast and identifying data gaps,...(and for) high-priority monitoring and research projects...over the next decade," explained Oman. "So that's where some of our research can come in, and (it) allows us to do some baseline work, too."

"The other component (of that $124,000 per year) is $74,000 for 10 years, and that's for Sara's program, allows her to get it off the ground, get the training underway (and provides) the administrative costs. We'll be able to do 14 training classes (annually) for two years." Going beyond those funds, Oman and LaBorde told how Tidewater Environmental Services of Vancouver, Washington, which ships petroleum products up and down the Columbia River, has set an example by taking responsibility for stewardship of the river. "They asked 'How can we help? Can we provide the safety training?' Following that lead, Shell Oil said, 'Okay, let's...talk about what we do out of Anacortes.' So, with the help of the industry, limited funds to carry on the training and planning will...go further," said Oman.

Added LaBorde, "It really is (a) perfect example (how) responsible government (programs) should be. It's government, citizen volunteers in training, and then it's industry (in support). And we're only going to get there if all three work together. If any one of those partners doesn't live up to their part in that triangle, we're not going to make it--well, we're not going to make what we could do. The biggest goal is going to be raising the money for that bird treatment center," she said in 1993.

A point of beginning to describe the spill response rescue and education work involving LaBorde is the Washington Wildlife Rescue Coalition, created by the Legislature but without funding to implement their plans. It was *Nestucca* money that targeted dollars to put plans into action when Oman came on board. Coalition members are Washington State Fish & Wildlife, Ecology, Adopt-a-Beach, CleanSound, U.S. Fish & Wildlife Service and the Washington Conservation Corps. The veterinarian is Jessica (Porter) Lainson; and Curt Clumpner, Northwest representative of International Bird Rescue, is the rehabilitation specialist. These two in particular, and all the people in the coalition brought valuable experience.

"It's incredible," says Sara. "I can't even put into words how lucky Washington is to have all these people develop its plans. We seem to have created a very effective system (that) will pay off in future spills, although there still will be the problems associated with emergency response. But whatever happens we'll know it would have been worse (without advance planning). All the coalition members are volunteering their time (outside full time jobs). They needed someone who could focus on this and get it done, somebody dedicated to setting up the system."

Basic strategy is simple, with but few changes from the process used starting with *Arco Anchorage*, then *Nestucca* and *Tenyo Maru*.

Collecting oiled birds on the beaches has changed little except for introduction of special boxes to carry different sizes of birds. But a primary care trailer now will operate as close to the beach as possible for temporary treatment to stabilize the oiled birds. Their eyes will be cleaned, they will be given special fluids for both nourishment and to minimize the effects of oil swallowed while preening, and they'll be kept warm. Once primary care has been given, the birds can be kept overnight and transported to a permanent bird treatment center. There they will be gone over with care, with each bird given a packed-cell blood count test. That test will reflect the condition of the bird, how well it is dealing with the stress and other aspects of being oiled, and whether it can live through the washing process. If the bird's general condition is not good, the decision is made then, and a merciful injection ends it. Only the stronger birds, those in better shape, are moved into the rehabilitation process.

Experts are learning, Sara says, that the amount of oil on a bird does not by itself indicate how well it will respond to treatment. Gone are the days when every bird is treated in the hope it can make it through to recovery. "We've learned...that with better preparation, equipment and training," says Mick Cope, "we're able to markedly improve survival rates."

"Triage work and blood tests, also," adds LaBorde. "It's a directed response, if you can get on oiled birds fast, a quick response. Get fluid in them, in a warmed area, stabilize them, then move them into good treatment. Release success has gone up markedly. All the more reason that we rely on...bringing in experienced contractors like International Bird Rescue. They've dealt with thousands of birds, and not many people have that kind of experience and knowledge."

By late 1993, two vehicles had been acquired for bird rescue. A large renovated van, dubbed "Big Red," was equipped to transport oiled birds from the primary care unit to the rehabilitation center. It has been succeeded by newer, specialized vans carrying all the equipment for beach rescue and bird transport.The other unit, a 40-foot trailer specially built for the state and equipped to be the primary care center, now has been assigned to Neah Bay and is waiting transport there for tribal operation in case of an oil spill. It was improved with fans and a heater to provide clean warm air.

A new primary care unit was being prepared in 1999, thanks in part to an estate that allowed the bird rehabilitation program to purchase an almost-new 53-foot Fruehauf truck trailer with sliding axle, air-ride suspension and chrome wheels for the available $14,000 rather than the $19,000 owed on the $28,000 unit by the estate. The deceased had been a veterinarian, and his sister wanted the trailer used for a purpose in keeping with his life interest. The older primary care trailer had been built on a lightweight mobile home frame designed for one-way transport to a permanent location, and was suitable for permanent assignment to Neah Bay as part of the *Tenyo Maru* settlement.

The new trailer will contain communication equipment in its administration work area, including multiple telephone lines and computer station. A middle area of the trailer will be for bird examination, medical treatment and food preparation, while the third trailer space will have racks for up to 250 birds, warm and well ventilated. The ventilation change came about because after *Tenyo Maru* a major disease mortality developed. Birds are susceptible to aspergillosis, so fans were installed to solve the problem.

The *Tenyo Maru* settlement also included $500,000 for the permanent center, Chris Lane explained. This represented a national precedent, a first instance in which federal representatives agreed to allow part of an oil spill response assessment to be employed for a wildlife rehabilitation center. It was qualified with a matching fund requirement and a three-year limit. Wildlife folk hope that businesses involved in oil transport, processing and sales will contribute toward the expected total of $1 million or more.

Creating the permanent rehabilitation center is by far preferable, it is agreed, to the problems and high costs of finding and creating a bird center from scratch after a spill has occurred, and the oiled victims are waiting in dismal condition for succor. After each past spill, that pattern has been repeated: *Arco Anchorage* at Port Angeles, *Nestucca* at Grays Harbor, the *Exxon Valdez* in Alaska, and the *Tenyo Maru* off Juan de Fuca (for which St. Edwards was used as the cleaning center).

"We still have the $240,000 from *Nestucca*, so we have a total of about $740,000 toward the new rehabilitation center," said Lane, who already has succeeded in funding the new trailer. Washington State University is interested in perhaps placing adjunct faculty at the new center and assigning student externs to work on bird research. The State of Oregon, which has relatively little invested in bird rescue/rehabilitation, also suggested that if the new Washington bird center could be located in Thurston County it might be used to support oil spill incidents in Oregon.

"That suggested the thought of moving the bird center south from the Seattle area," Lane said. "We had to look at a variety of things such as water and sewage, so I went to LOTT (Lacey, Olympia,

Tumwater and Thurston County Wastewater Management Partnership) and asked them what is going on in the near future. They jumped on...right now through discussion of the potential to share in their master plan for satellite plants in outlying areas. They believe they'll reach capacity in the year 2002, and rather than expand the facility in downtown Olympia they want to build small plants in various locations that can take part of the load from the cenral plant."

Mike Sharar, LOTT partnership administrator, explained the concept of satellite treatment plants in more detail. Three such plants are contemplated: one in the Hawks Prairie area west of the Nisqually River, one south of Tumwater, and the third south of Lacey. Each would handle a million or more gallons per day of sewage that otherwise would flow to the downtown Olympia treatment plant, process it and return the "solids" (still 95% liquid) into the system's pipeline to the main plant. The remaining water at the satellite plant, after purification and chlorination, then will circulate through a series of "polishing ponds" until reaching the class-A purity standard suitable for any use except drinking. Finally, the class-A water will be allowed to drain through the layers of gravel and sand deposited thousands of years ago by glaciers, and so recharge the underlying groundwater.

The millions of gallons of water separated by each satellite plant, the LOTT administrator pointed out, would reduce the load for the main plant, decreasing wastewater discharge into Puget Sound. Priorities of this plan, Sharar explained, maximize utilization of the main LOTT treatment plant. By delaying the necessity for major expansion of the central plant, satellites will help accomodate planned growth, yielding maximum benefits to the environment by cleaning and restoring wastewater while producing multiple benefits for the community, along with equitable distribution of controlled costs between current ratepayers and new development.

When the bird rehabilitation center is constructed in the same vicinity as a satellite sewage treatment plant, it would employ class-A treated water for cleaning birds. Afterward the lightly oiled and soapy water would be sent back through the satellite plant for purification a second time and then into the polishing ponds. Additional ponds using class-A water might also be built for other purposes, including public swimming, catch-and-release fishing or raising salmon fry. The surroundings could become a public park.

Lane has a vision that would broaden the usefulness of the bird rehabilitation facility. He'd like it to be used as a learning center for schoolchildren to become familiar with seabirds and fisheries, sea mammal lifestyles, migration and habitats; to watch demonstrations on oil recovery, transport, refining, and how society is working to prevent oil spills and preparing for when spills do occur. The program would help teach young people to become good stewards for our total environment.

LaBorde suggests that tying together the LOTT satellite plant with the bird rehabilitation center has a special aspect. She recalls when St. Edwards was used for bird rehabilitation following the *Tenyo Maru* spill. "We dumped I think a million gallons of water down the drain (after bird washing). The power here is that the water won't be wasted, it will be recycled. It just has a nicer feel. To me it -- bird washing -- had a weakness because of the amount of water it took. This plan is a home run."

Financing construction of a permanent rehabilitation center seemed not too difficult as the Wildlife trio discussed the situation. The *Tenyo Maru* $500,000 will have to be matched dollar for dollar within the next three years, Lane explained.

"I think the (oil) companies are very interested," said LaBorde, "that they'll come to the table and go with your plan. It makes good sense for them. It's our responsibility to see the project is done well. We've stepped up and coordinated Wildlife response to spills at the state level, but it's oil companies' responsibility to make sure they're prepared to deal with it if they have a spill. Of course nobody is going to walk away from somebody putting money on the table."

Turning to the need for training volunteer workers to collect oiled birds from the beaches, help operate the primary care trailer and then staff the cleaning and longer-term care at the rehabilitation center, Lane is seeking a financial grant. The training program, in abeyance for a few years, will be reduced from eight hours to four, setting aside the hazmat (hazardous materials) aspect and focus on wildlife rescue. He intends to concentrate first on Thurston County volunteers to staff the center, assuming location plans work out. Training for beach workers, also a necessity, will include collection of dead birds, properly labeled for use as evidence in court. Beach worker classes could be conducted at the Fish & Wildlife regional office in Montesano, or in various other locales closer to possible oil spill areas, including Clark County and Seattle, etc.

"I concluded that training is useful for two purposes," explains Lane. "Training folk who might be available during a spill will also be good public relations to help get them involved. The question remaining was, 'how many would realistically be available following a spill?' Training is expensive, which was my reason for eliminating the hazmat part. So now a volunteer can come in, get a brief refresher, start volunteer work and we can give the hazmat piece on-site. It's cost effective, training on the job."

LaBorde in 1993 had been working to broaden the detailed information in the geographic response plans (GRPs) with the help of residents. The local Department of Emergency Management agent and other interested or concerned people are gathered, so that "It's not just someone from Olympia coming in, not at all. You can't just bring in people from somewhere else and expect plans to work smoothly. Local search and rescue teams, for instance, can both help and be helped by joining these planning sessions. We're working now with Clallam County. People there develop data sheets by going along the beaches marking all the access points on their detailed maps, noting such items as viewpoints from which a person can see all of a beach. They also check the accuracy of GRP maps as to which beaches are sandy, cobble, rocky or have inaccessible cliffs..., and places where debris collects. And they will provide details such as 'This is how you get there, but in February that road's probably washed out,' or 'This place you have to reach, but it's private land and the only way to get access is to call Joe Joseph.' We need somebody to know how to reach Joe.

"And that regional/local group plan," Sara continued, "is entered into our computer system. Then, if we have a spill or some other emergency, the computer will show all the details. Local contacts also will be included, in case of special equipment needs, or perhaps a shortage of volunteers. We want to have a local group available to help set up and staff the primary care station, giving that support. We don't want to upset the local system, but work with it.

"A woman called from the Longview RSVP (Retired Senior Volunteer Persons), and didn't know we had a system in place. Finally, through about 15 other people, she reached me to talk about setting up volunteers to help in an emergency. I told her 'What really would help is transportation between Cathlamet and Longview. I have veterinarians on call if we have oiled birds and can take them to HOWL Wildlife Rehabilitation Center.' She said, 'I can take care of that for you. We can set up a transport network.' Now I have a map of where all the RSVP groups are, and we can tie in with them in each local area, and with other groups that are already there. Most people volunteer to do something, not to wait, and we want to tie in with local groups that already exist and can help during a spill. We also want local people trained with the required basic safety course, because they can't work unless they've had that. Then we'll go on to leadership classes to create a nucleus of people who can supervise untrained volunteers that converge to an area to help in an emergency," said LaBorde.

Less visible but also important to the advance preparations is the pre-recorded menu-driven telephone setup, 1-800-22BIRDS. It can be activated when a spill occurs, provide basic information, and enable callers to reach an operator.

The goal of the Wildlife Rescue team is to be able to set up search and collection, along with the primary care unit, within the first 12 hours of a spill. Within 24 hours the aim is to have the rehabilitation care center ready to go. They're relying on the Washington Conservation Corps, which helped immensely on *Nestucca*, and Ecology teams to help with search and rescue, and with setting up the primary care center.

As we concluded our 1993 interview, LaBorde added some final observations. "I don't like telling anybody that we can deliver unless we can. Now...we're ready to put this show on the road and explain what we're doing and why. The general public cares about this, and...they feel frustrated when a spill occurs. It's a natural tendency. We feel for those victims,...want to respond. So now we have a system that...can be effective using volunteers...safely...to help oiled wildlife. Taking care of this problem (will mean) the Coast Guard doesn't have to deal with it, nor NOAA, nor CleanSound or others working on cleanup. They can depend on us to respond to oiled wildlife while they concentrate on the oil and resource protection." That was still a good summary in 1999.

The *Nestucca* settlement also included another 10-year annual $26,000 payment to Ecology for the natural resource damage assessment (NRDA) program chaired by Dick Logan, and single payments of $43,000 for Ecology's damage assessment work, plus reimbursement costs for all agency response that came to $647,000.

Oman, assigned to the Natural Resource Damage Assessment interagency group at the time in 1993, asserted that it is one of the biggest things from the *Nestucca* legacy. "It's all the natural resource agencies in the state: Ecology, Fish & Wildlife, Natural Resources (forestlands), Parks & Recreation, Archeology & Historic Preservation and Health (Office of Shellfish Programs). We have several functions," she explains. "One is responding to spills and coordinating there. And we talk about what we do, how we do it, why we do it, so that we (avoid) duplication, and are filling in the gaps so that we have tight interface. We also work on damage assessment, identifying injuries from a spill and putting a dollar value on the injury and recovery response. And we do restoration projects. As the costs (dollar assessments) come in as they have...the last few months, we'll do more restoration projects.

"Natural resource agencies never before have had appropriate funding to do this," she explained. "I think that for outsiders there is an opinion that it is just very easy as part of what we do in our other activities. But there are limitations on people's time and availability when you get involved with a comprehensive project such as oil spill response. It's just wonderful to be given the opportunity...finally (to) pro-actively do something to plan for the next event, so as to be able to respond effectively and efficiently.

"Now the responsible party in a spill is billable for all activity from response through assessment and restoration," she continued, "so we can get reimbursement. That leaves only the costs for small spills and for unidentified spillers coming out of the state pocket.

Dick Logan, NRDA chair, is just as positive. "In 29 years in fish and wildlife agencies, of which I spent about 15 in Alaska, it's the first time I ever saw a fund that came back to doing what it was supposed to do, and not just being dissipated somewhere else," said Logan. "The beauty of NRDA is that...you're going to get bucks back (after a spill), and the NRDA team then becomes the focal point for disbursement of these funds. This same set of players...can suggest studies,...can suggest restoration, because the whole idea is 'you restore what you lost.' We have to deposit (these dollars) in what's called the Coastal Protection Fund. The beauty of that, the reason we were able to sell this whole side of the concept was that money can only be invested in basically restoration, and you can't fritter it away on hiring people and doing bureaucratic things. It has to come back to this business.

"I was intrigued. That was one of the reasons I signed on to work on this, it intrigued me that we could actually invest back in resources," said Logan.

Explaining the background that set the stage for NRDA, Logan said, "We use the CERCLA regulations as the appropriate vehicle to do damage assessments. (CERCLA refers to the U.S. Comprehensive Environmental Response, Compensation & Liability Act of 1980.) We will now have to use NOAA on oil spills and CERCLA on hazardous waste sites. The reason you have to use those federal regulations to do a formal damage assessment is (they) give you what's called a rebuttable presumption (in a court of law) so that they've got to prove you wrong.

"We found out real fast, it's awfully hard to quantify those uses or losses that you can't directly measure. It becomes very expensive.

The way the NRDA team works now is that after a spill, and regardless whether it was oil or another hazardous substance, or whether it occurred on fresh water or in the marine environment, "We all get together in a room,...now we have a...choice. We can either...do a formal resource damage assessment...under the federal regulations...or use our state compensation schedule regulations. In the last 15 months we've had 1,624 reports of oil spills in this state. We never have gone the formal NRDA route on any of these.

"What it (the compensation table) does is, it looks at environmental sensitivity, at oil toxicity, and...at actions of the RP (responsible party), which is, did they bust their butt to take care of it or didn't they. Our state setup also allows the 'feds' to come in on it, the tribes (or anybody defined as a trustee), and it allows for the RP to play a role in the process.

"So this team collectively comes together and makes the decisions," Logan explained. "Generally we have to act as a team. In the past, this has run lots of different ways. On a nationwide level, it's sort of chaotic, and so this year (July, 1993) Vice President Al Gore appointed eight of us nationwide to redo the national process. Each state has (and)...most federal agencies have their own process. Protection of 'turf' and endless red tape has in some instances resulted in more...being spent on assessment than is recovered for restoration. Each assessment takes years to complete.

"It came out real fast in that national meeting," Logan recalled, "that there are two or three states in the lead, and Washington is one of them. Florida has been pushing real hard,.... making up a compensation schedule...with a slightly different twist. Texas has been active, and California has picked up because they have a barrel tax (similar to Washington's). The rest (of the states) are at all levels in activity."

The proposed national plan is based on Washington State's Pre-assessment Screening and Oil Spill Compensation Schedule, combined with coordinating teams of state and national agency officials concerned with the resources affected by a spill. Early coordination is planned to eliminate redundancy between agencies and to require open dialogue and early information flow. Once finalized by the National Performance Review Task Force, it will be incorporated into a pilot program. "It gives grassroots teams the responsibility to design the system," said Logan. "Since the team knows it is responsible from start to finish, participants are more likely to take pride in their work and their finished product."

To further one's understanding of the oil spill compensation schedule, excerpts from the 24-page abstract (15-7) may be helpful. "The compensation schedule for marine and estuarine environments consists of two main components: the resource vulnerability...and the oil effects.... The resource vulnerability rankings collectively rate the sensitivity of the receiving environment to spilled oil, whereas the oil effects rankings rate the propensity of a spilled oil to cause three types of adverse effects: acute toxicity, mechanical injury and environmental persistence. Both...were constructed using the best available scientific information and expert opinion.

"The resource vulnerability rankings incorporate resource, area and seasonal specificity to derive a composite resource vulnerability score for a particular spill. To incorporate area specificity, the marine and estuarine waters of Washington State were divided into 131 subregions (The Columbia River...is treated...separately). Resource and seasonal specificity were taken into consideration by independently evaluating the oil spill vulnerability of several resource categories on a subregional and seasonal basis. ...Resources are divided into...seven categories: habitat, marine birds, marine fisheries, shellfish, salmon, marine mammals and recreation. ...Rankings (were) developed in consultation with an advisory committee...of resource experts from state and federal agencies, academic institutions, consulting firms, Indian tribes and environmental organizations. All...rankings utilize a one-to-five rating scale, where...five represents the most vulnerable...and...one...the least vulnerable condition."

To observe how the NRDA team operates, I sat in on the November 10, 1993, session hearing the Pre-assessment screening report on a spill October 14 at the Tidewater Barge Lines terminal on the Snake River near Pasco. Also present was Greg Yaroch, manager of Tidewater Environmental Services, representing the barge line. Yaroch, a former Coast Guard officer, had been USCG on-scene representative at LaPush during *Nestucca*.

Rob Whitlam, representative to NRDA from the Department of Community Development's Office of Archeology & Historic Preservation, gave the oral report based on information compiled by Ecology responders Don Beery and Jeff Dill. An estimated 3,295 gallons of diesel oil escaped from a 4-inch slit in the bottom of a Tidewater barge moored at the Snake River terminal, which is two miles upstream from the Snake's confluence with the Columbia River. The spill was discovered about 7:15 a.m. and reported to Ecology an hour later, after preliminary response measures had been taken.

Officers from Ecology, Fisheries and Wildlife flew over at 11:30 a.m. and "spotted...patches of oil for several miles upstream of McNary Dam. Oil was not seen again until just downstream from the Snake/Columbia River confluence. Heaviest oiling was...along the north bank of the Snake River from the terminal (downstream) to Sacajawea State Park. Tidewater Environmental had deployed boom around the barge, downstream at the Chevron Terminal and at (the) state park. Booming was not always effective in containing the oil due to delays in changing boom configuration as winds shifted. While large flocks of waterfowl were seen in the area near the McNary Refuge and nearby islands, they were not seen in the immediate vicinity of the oil."

Yaroch, on behalf of the RP, provided information that had not been available the day of the spill. Cause of the spill was a large boulder on the bottom of the river 200 hundred yards downstream, which he said was discovered by divers. They had inspected the crack in the bottom of the barge and found a long scratch/indentation leading to it. When they found the boulder it carried evidence of having scraped the barge. Except for the rock, the barge with its draft of 14 feet 6 inches would not have touched bottom as it was taken upstream. Divers moved that rock and others nearby to deeper water.

Oil did not leak from the barge immediately, Yaroch said, because a "water blanket" developed at the site of the crack. During the night after the barge was delivered to the terminal by tug, oil was pumped ashore, causing the barge to shift its stance, and oil then began to leak.

The first containment boom was deployed at 7:30 a.m., only 15 minutes after discovery of the leak, and two more booms were deployed before 7:50, one at the terminal and another downstream at the Chevron terminal.

Yaroch reported that at several skimming sites vacuum trucks recovered a total of 2,260 gallons of the spilled fuel, mostly at the Tidewater dock. That, he noted, was 68 percent of the total spill. Asked by a member of the NRDA team concerning the high rate of recovery, Yaroch said, "I attribute it to your requirements, to our preparation, practice and prompt response under the best possible conditions." And he added a significant observation relative to preparation and training for oil spills: "This is the payoff."

At the conclusion of the discussion, all five NRDA representatives voted against instituting a formal Resource Damage Assessment process, and also against a formal restoration process, since there was no known substantial resource damage. The only evidence of that sort was the report of a beaver at Sacajawea State Park suspected to have a small patch of oil on its fur. Therefore, the standard rules will be applied under the compensation schedule worksheet, with credit given for quick and effective response. Yaroch inquired whether recovered oil is deducted from the quantity spilled, and the response was "no, because it has been in the water."

The final NRDA assessment was $38,058.32 as a result of the Tidewater Environmental spill. That Tidewater decision presumably took into consideration the prompt and effective response, and compares with the charges assessed the operators of the *Nosac Forest*, a 598-foot cargo vessel operated by Barber International of Oslo, Norway. That firm had to pay $122,696 for environmental damages from a 6,260-gallon spill April 21, 1993, in Tacoma's Blair waterway (15-8). That was the first NRDA assessment under the state's new compensation schedule. Ecology had previously fined the company $90,000 for the crew's failure to exercise reasonable care during fuel transfer, causing the spill.

The largest NRDA assessment was the $5.2 million *Tenyo Maru* settlement, Logan explained in late 1999, and the smallest was $97 for a 25-gallon spill. Out of 190 cases handled by the state damage assessment process, only about a half dozen have not paid, and those cases involved miscreant foreign firms that left American jurisdiction. The money collected through smaller judgements, Logan said, is collected until at least $100,000 is available in the account, then the NRDA committee chooses among rehabilitation needs for effective utilization. Spartina grass elimination in northern Puget Sound and on Willapa Harbor is one example, along with salmon stream rejuvenation and other wildlife restoration projects.

This story might have ended here except for other developments. One problem is financial, relating to the barrel tax on oil transported into the state, and brought about a citizen initiative in 1996 aimed at broadening the 5-cents per barrel tax, which had proved of limited effectiveness because of built-in rebates. Although well-intentioned, it failed of passage, perhaps in part because of excessive breadth in an attempt to solve a variety of problems. Limited funding under the barrel tax continues to hinder state response programs.

Observers might suppose that increased awareness, prevention, training, planning, new equipment and preparation have been effective. But don't count on those efforts preventing disaster, though they'll help.

Tankers haul North Slope crude from Valdez, a deep-laden parade across the storm-swept Gulf of Alaska to Puget Sound, with ships older and more vulnerable each passing year. Oil and other fuels are also processed and transported in every direction by ship, barge, rail, truck and pipeline throughout Washington and neighboring jurisdictions. Volume far surpasses public awareness.

Problems, however, have increased rather than decreasing. For instance, the Ecology Spills Program has struggled with the increasing plague of illegal methamphetamine labs and its resultant backroad dumps. The workload keeps growing, says spill response supervisor Steve Hunter. As the lab numbers continue to soar, the spill response budget has remained at its 1994 level, which was intended to cover 30-50 labs per year. In just the first seven months of 1999, Ecology teams responded to 381 labs and dumps (15-9). Federal assistance is being sought.

No major marine oil spills occurred in the Pacific Northwest for several years, until February of 1999. The freighter *New Carissa* grounded on the ocean beach just outside Coos Bay, Oregon, while waiting to enter the harbor to load a cargo of wood chips. After a few days of pounding in the surf, bunker fuel began to leak; and when an incoming storm threatened the vessel, its fuel tanks were set afire by Navy demolition experts. Much of the fuel was burned, but the hulk broke apart and thousands of gallons escaped (15-10).

The forward section was finally pulled off the beach, but heavy weather broke the towing cable and grounded the partial hulk again at Waldport, eighty miles further north. On a second try with heavier cable, it was successfully towed to sea to be sunk by naval gunfire and a final torpedo. The stern section, though partially dismantled, was still on the beach in late October after further unsuccessful attempts to pull it off. It was left until spring and possible demolition by crews using acetylene torches (15-11). The *New Carissa* was given considerable news coverage over nine months, and cost its owners and insurors up to $35 million (15-12), but its oil spill was limited and cleaned up by response crews.

"Statistically, pipelines are the safest" way to move petroleum products, according to Garry Schwan, director of commercial policy for ARCO, part owner of Olympic Pipe Line Co., which reaches from refineries in northwest Washington south to Seattle, its airport, and on to Portland, Oregon (15-13).

That may be correct; but Olympic's pipeline rupture in June, 1999, allowed 277,200 gallons of gasoline to pour into Bellingham's Whatcom Creek and flow through a major city park (15-14). It burst into flame when two youngsters unknowingly lit a match. Both died of massive burns, and a trout fisherman drowned after being overcome by the fumes.

Had the gasoline flow continued for only a minute or so longer, the explosion and fire could have wiped out much of downtown Bellingham. Even so, a mile and a half of park along the stream's course through a canyon was burned to devastation, including all water life and many acres of trees and undergrowth. The investigation that followed, still uncompleted by the thinly-staffed federal regulatory agency, brought complaints that were carried back to a congressional hearing. Pre-disaster Olympic Pipe Line information secrecy, however, also went up in smoke, amid promises to pay for damages and prevent future such disasters.

News reports revealed inexcusable influence by industry to weaken pipeline safety regulations (15-15). Not only were ex-bureaucrats becoming lobbyists for the pipeline industry, but a decade of pipeline accidents in the United States has killed 56 people, injured more than 219 others, destroyed homes and caused millions of dollars in damages. Most were the result of carelessness or lack of supervision. Yet efforts to tighten regulations have resulted instead in relaxation. "The industry really gets to set its own performance standards" said Susan Harper, executive director of Cascade Columbia Alliance (15-16). Officials in various levels of local and regional government in Washington State now have joined forces with New Jersey, Florida, Virginia and other victimized locales to renew the campaign for more effective controls. One troublesome aspect has been the lack of resources to perform regulatory inspections. On the state level, "the law is in place but it can't be enforced," said Mark Asmundson, mayor of Bellingham (15-17). He chaired a seven-member Fuel Accident Prevention and Response Team created by Governor Locke to prepare pipeline safety recommend-ations. Leading some three-dozen suggestions was creation of a strong state office of pipeline safety, if financing and authority can be made available (15-18). "Nothing can make up for what's happened," said Asmundson, "but many serious people have worked very hard to accomplish the goal that this will never happen again."

Olympic Pipe Line Co. made news again when its president, Fred Crognale, sought to delay testifying in the civil suit initiated by families of the Bellingham victims. Crognale and other company officers faced possible criminal indictment by a federal grand jury in Seattle as a result of investigation into the June explosion (15-19). And that same day of December 1999, the company was assessed a $120,000 fine by the Washington State Dept. of Ecology for negligence involving an August 29 spill of 3,300 gallons of mixed fuels at its pumping station in Renton. Steve Hunter, spill response manager at Ecology, said, "We thought after the Bellingham tragedy, we would find...improvements." Instead, regulators found "a pattern here that is alarming." He termed the fine "escalation,...stepping up the enforcement pattern." In addition, Olympic was given a timeline in which to complete needed improvements, including maintenance and spill response plans (15-20).

All the precautions, accumulated skills and knowledge cannot prevent the human or mechanical error that someday, inevitably, will precipitate the next major oil spill, marine or otherwise. There is indeed higher response capability than in prior years. Yet spilled oil, when it happens, is just as dangerous as before; and our fish, birds, other wildlife and lesser organisms are depleted and more vulnerable. Oil spills reveal weaknesses, and even our improvements are relatively untested.

Ecological awareness has increased, but we live with only remnants of the original natural community, unable to replace what has been squandered or lost. Strengthening our sense of community, and creating new elements of responsibility and sharing, is a process. Caring for the earth, our home, depends on working and learning together. We learn from what oiled birds teach us.

Anybody who has experienced an oil spill, even the relatively small but widespread *Nestucca* event, and has seen photographs of much greater disasters such as in Alaska or those in the Atlantic, knows it can happen along any coast. We've smelled one spill that we'll never forget; and we can, unfortunately, "smell" the next one coming.

That has been the message of *Nestucca*, to all who share or inherit these concerns.

Remembering the Nestucca People

Curiosity is educational, notwithstanding the kind that killed the cat. It's one quality required of news reporters, and is part of the reason newspapers and other media exist. My story of the *Nestucca* and its spawn now is finished, but not complete. It lacks antecedents.

When I asked the meaning of "Nestucca," the only answers were that it is the name of a river on the Oregon coast and an Indian tribe that once lived there. As I pursued information for the *Nestucca* story, that small mystery continued to puzzle. Finally, through the Washington State Library, I got in touch with Wayne Suttles and Volume 7 (which he edited and partly wrote) of the *Handbook of North American Indians* (16-1). Suttles referred me to William R. Seaburg, author with Jay Miller of the chapter on the Tillamook tribe of which the Nestucca people were the southern band.

The library also had a small volume, *Tillamook Indians of the Oregon Coast* (16-2), by John Sauter and Bruce Johnson, a pair of curiosity-driven educators. Their work, which included thorough research and even careful excavation of ancient village sites, has brought to light some of the Tillamook people's history and knowledge of their society that had been hidden by time. Some of their stories are gems.

One Nestucca village was near the present site of Pacific City, Oregon. Another was on the inshore side of the Nestucca sand spit, thought to have been a seasonal settlement "both for the Nestucca band of Tillamooks and for other tribes from the Willamette Valley who visited to share the salmon harvest."

The Nestucca people, however, have vanished from their native lands and waters. Like other first people along the Pacific Coast and elsewhere, they had no resistance to such European-borne diseases as smallpox, syphilis, influenza and tuberculosis. Sauter and Johnson (op. cit. 16-2) report that in 1788 Captain Robert Gray "noted pockmarked faces on a number of the natives at Tillamook Bay--results of smallpox. In 1806, Chief Concomly of the Chinooks told Captains Lewis and Clark of a great epidemic...in 1802, four years prior to their wintering at Fort Clatsop. The captains estimated the Tillamook population at 2,200; yet by 1845 the Wilkes Exploring Expedition judged it to be only 400. In 1851, when Warren Vaughn settled on Tillamook Bay, there were fewer than 200. By 1871, according to the 'Inventory of the County Archives of Oregon,' there were only 28 Nehalems, 83 Tillamooks and 55 Nestuccas."

An intriguing, indeed shocking, proposition is put forward as a contributing factor to deaths from smallpox. "They chose an unusual cure. The afflicted would enter a sweat-house for a long period; then, at the height of his sweat, he would plunge into an icy stream, hoping to kill the disease spirit. This, in fact, did kill the spirit, but it took the body with it."

Helped along by alcohol and tuberculosis, introduced by American and European explorers and pioneers (invaders), the Nestucca and Tillamook people are gone, though related folk survive along the Oregon coast.

The outstanding legacy of the Nestucca people is their relationship to salmon, and the fish themselves. Although not the untold thousands that swarmed up the Nestucca and Little Nestucca rivers 200 years ago before new arriving settlers took over, there are sizable migrant and resident populations of both salmon and trout. A fisheries biologist at the Oregon Fish & Wildlife office in Florence said there are both spring and fall chinook, coho and chum salmon, summer and winter steelhead, sea-run cutthroat trout, resident cutthroat and even white sturgeon on the Nestucca River. The Little Nestucca also has all those listed for the Nestucca except spring chinook and steelhead. The Nestucca's fall chinook run was termed healthy, the others less robust. Counts of spawning salmon averaged 123 fish per mile in 1992, but fell to 94 per mile in 1993 when low stream levels for lack of rain caused more chinook to spawn in the lower river.

Chinook salmon are mainstream spawners whose fry travel to salt water rather than growing for a time in the stream as do the young of other migrant salmon or the trout, so are less exposed to stream-side logging, urban development and pollution. Catch records on fall chinook by sports fishermen have been 2,200 to 2,300 fish in recent years, varying from a low of 1,479 to a high of 3,700 over a longer period. Fall chinook are about 85 percent native fish, and the spring chinook run is about half natural and half hatchery fish.

Back, now, to the original question: What is the meaning of Nestucca? Quite simply, it is Salmon River (16-3). The natives called themselves Staga-ush (Stucca-ush is an alternate spelling) or "people of the salmon." Their culture revered the salmon, and they harvested only what they could use in a season (16-4).

Since the 1970s and the so-called (U.S. District Court Judge George) Boldt decision (16-5) awarding half the salmon to treaty tribes, first Americans have been restored to their rightful place honoring and maintaining the traditions with responsible care for the heritage of all wildlife and other natural resources. The rest of us, who arrived less than 300 years ago and are heirs to more than a century of wasteful exploitation, can and should honor and work with the children of native folk who have lived here untold thousands of years.

Our debt to those first people, and the natural resources on which they depended for food and all aspects of life, can only be met by responsible caretaking. Only then will preservation continue for all, both their children and our children together.

Endnotes

Chapter 1

1. *The Daily World*, Aberdeen, Washington, March 12, 13, 16, 17, 1964.

2. *The Daily World*, Aberdeen, Washington, March 27, 1964.

3. State Pollution Control Commission minutes, April 7, 1964, Washington State Archives.

4. *Seattle Times*, Jan. 8 and 11, 1972.

5. *Journal of the Fisheries Resource Board of Canada*, Vol. 35, 1978.

6. *Seattle Times*, Nov. 14, 1989, and *The Olympian*, Dec. 12, 1990. Bird loss verified by USF&WS; also by Wa. State Dept. of Ecology, Marine Resource Damage Assessment Report 87-4, January 1987.

7. *Associated Press*, Dec. 12, 1990, sundry papers.

8. *Seattle Times*, Nov. 14, 1989.

9. *46 USCS* (May, 1993 Cum. Supp., *United States Code Service, Lawyers Edition*, Lawyers Cooperative Publishing, Rochester, N.Y.) par. 3703a Tank Vessel Construction Standards, sec. (b)(3). New tank vessels constructed after January 1, 1994, must have double hulls; but a retirement schedule is included in the law for existing vessels so that a 29,000 gross ton tanker with single hull (for example) can continue to operate until it becomes 25 years old Jan. 1, 2005. If a similar vessel turns 40 years of age Jan. 1, 1995, it can no longer operate on U.S. waters.

10. Sundry news stories, 1988, and Wa. State Ecology records.

Chapter 2

1. Report (15732/001POR89/MC89001217 dated March 20, 1989) to Commandant (G-MMI-3) from Investigating Officer Lt. M.L. Emge via: (1) Officer in Charge Marine Inspection, Portland, Oregon, and (2) Commander, Thirteenth Coast Guard District (m). Subj. M/V *Ocean Service*, O.N. 507101, collision with T/B *Nestucca*, O.N. 569658, in the Pacific Ocean at position 46 degrees 54 minutes 55 seconds N,

124 degrees, 14 minutes, 52 seconds W on 22 December 1988, with no personnel injury.

2. In these extensive quotations from the depositions, some questions and answers have been moved or combined for clarity and continuity. Various attorneys were involved in the questioning, representing Sause Brothers Ocean Towing, the State of Washington, the United States, Canada and British Columbia. All were parties to the U.S. District Court action at the time, in 1990, preparatory to the trial on liability that December. The U.S. and Washington State reached an out-of-court settlement prior to trial. Other parties have settled since.

3. *Seattle Post-Intelligencer*, January 10, 1989.

Chapter 3

1. "*Nestucca* Oil Spill On-Scene Coordinator's Report," page 6, August, 1989, Washington State Department of Ecology, Christine Gregoire, Director; Lewey Kittle, State OSC.

2. *The Olympian*, Olympia, Washington, January 8, 1989.

3. "*Nestucca* Oil Spill Report" page 9, June 1989, prepared by Canadian Coast Guard Western Region, Victoria, B.C.

Chapter 4

1. *Quinault Natural Resources*, "The December Oil Spill" by Jacqueline M. Storm, QIN; winter/spring, 1989, a publication of the Quinault Nation, Taholah, Washington.

2. *Seattle Post-Intelligencer*, January 5, 1989.

3. *The Olympian*, Olympia, Washington, January 8, 1989.

4. *Shipwrecks Off Juan de Fuca*, by James A. Gibbs (Binfords & Mort, 1968) is at variance with the contention that oil could not travel eastward to Destruction Island. After the 1875 sinking of the steamer *Pacific*, according to Gibbs, bodies of drowned victims were found as far as 150 miles eastward on San Juan Island. Gibbs explains on his page 40 that the incoming current in the strait is much stronger than the outgoing and "a large part of the water... passes out through the Gulf of Georgia and the channel east of Vancouver Island."

5. *Quinault Natural Resources*, cited above; Workman agrees with Oberlander, "There was a lot of oil, but 'knee deep' I think was just Jim's phrase."

Chapter 5

1. Diane Harvester's notebook/diary used with her permission, and with the cooperation of Washington State Department of Ecology and Washington State Archives, the diary custodian.

Chapter 6

1. *The Daily World*, Aberdeen, Washington, December 23, 1988.
2. *The Daily World*, December 24, 1988.
3. *The Daily World*, December 26. 1988.
4. *The Seattle Times*, December 25, 1988.
5. *The Seattle Times*, Jan. 1, 1989. See also *Northwest Naturalist*, 71:88-92, Winter 1990, "Some Effects of a Major Oil Spill on Wintering Shorebirds at Grays Harbor, Washington," by Eric M. Larsen and Scott A. Richardson.
6. *The Daily World*, Aberdeen, Washington, December 28, 1988.
7. *The Daily World*, Aberdeen, January 15, 1989.
8. *The Daily World*, Aberdeen, December 30, 1988.
9. *The Daily World*, Aberdeen, January 4, 1989.
10. *Peninsula Daily News*, Port Angeles, January 3, 1989.
11. *Associated Press*, January 1, 1989, various papers.
12. *Seattle Post-Intelligencer*, January 6, 1989.
13. *The Herald*, Everett, Washington, January 15, 1989.
14. *The Daily World*, Aberdeen, Washington, February 9, 1989.

Chapter 7

1. Instructions for tube feeding birds and other procedures are fully covered by International Bird Rescue's *Rehabilitating Oiled Sea Birds: A Field Manual*, published in 1985 by the American Petroleum Institute, Washington, D.C., as API publication 4407. It was based on a 1983 manuscript by Anne S. Williams, D.V.M., and edited by J. Burridge and M. Kane of API. Also refer to the interview with Alice Berkner, IBR executive director, in Chapter 6 this volume, for more recent information on water quality.

Chapter 8

1. "One on One" by Constance Perenyi, excerpted with permission. This article first appeared in *The Living Bird Quarterly*, Vol. 9, No. 1, Winter 1990. *TLBQ* is a publication of Cornell Laboratory of Ornithology, Ithaca, N.Y.

2. *Willapa Harbor Herald*, Raymond, Washington, January 11, 1989.

3. *Northwestern Naturalist*, Winter 1990 (published April 1991), "Some Effects of a Major Oil Spill on Wintering Shorebirds at Grays Harbor, Washington," by Eric M. Larsen and Scott A. Richardson; quoted with their permission.

4. "Seabirds in Washington Offshore Zone," Terence Wahl, *Western Birds*. Vol. 6, Nov. 4, 1975; also "Distribution & Densities of Marine Birds on the Canadian West Coast," March 1980 report to Environment Canada by K. Vermeer, M. Lemon, I. Robertson, R.W. Campbell, G. Kaiser; Audubon Christmas counts, *American Bird* vols. 34-43, 1980-1989, show 1985-6 count at Grays Harbor recorded 8,000 murres and the 1988-9 count 10,000 murres.

5. "Seabird Mortality Resulting from the *Nestucca* Oil Spill Incident, Winter 1988-89," prepared for Washington State Department of Wildlife by Ecological Consulting, Inc., Portland, Ore., submitted June 1991.

6. "Washington Coastal Murre Fluctuation Related to *El Nino*," by Ulrich Wilson, *The Condor*, 93:853-858. Mr. Wilson has also provided USF&WS counts obtained during multiple murre surveys 1995-1999 which demonstrate drastic decline of murres the final year. Highest counts were 6,700 in the spring of 1995 and 7,500 in autumn of 1996. The four 1999 counts were about 3,900, 1,000, 2,000 and a final tally fewer than 200. Possible reasons for near elimination of murres include a food supply decrease related to ocean current warming, reported from British Columbia biologists in an Associated Press story from Vancouver, Oct. 29, 1999. It quoted David Welch, Canadian Fisheries Department scientist: "We have a very serious, big-picture problem," relative to high seas salmon. "It's worse than people think." Warming has already cut productivity of British Columbia's offshore waters 40 to 50 percent over the past decade, said Frank Whitney, chemist with the Institute of Ocean Sciences. Obviously, one expects marine bird food supply also to be affected.

Chapter 9

1. Excerpted with permission from "Silent Shores" by (William) Randy Thomas, Feb. 16-22, 1989, *Monday Magazine*, Victoria, B.C.

2. *Coastal Oceanography of Washington and Oregon*, 1989, edit. by Michael R. Landry and Barbara M. Hickey; Elsevier Oceanography Series, 47; page 44.

Chapter 10

1. *Guide for Selection of Tankers*, Tanker Advisory Center, Inc., 217 East 85th St., Suite 259, New York, N.Y. 10028.

2. *The Columbian*, Vancouver, Washington, Jan. 8, 1989.

3. "*1988 Petroleum Estimates for Puget Sound & the Strait of Juan de Fuca*," by Scott H. Chadbourne & Thomas J. Leschine, Institute for Marine Studies, University of Washington, Seattle, Wa. 98195.

4. Volpe Sudy: *Scoping Risk Assessment; Protection Against Oil Spills in the Marine Waters of Northwest Washington State.* Prepared by the Environmental Engineering Division, Office of Systems Engineering, John Volpe National Transportation Systems Center, 55 Broadway, Kendall Square, Cambridge, Massachusetts 02142-1093; prepared for Human Element & Ship Design Div. (GMSE-1), U.S. Coast Guard Headquarters, Washington, D.C. 20593.

5. *46 USCS 3703*, op. cit., page 16 (Act Aug. 18, 1990, P.L.101-380, Title IV, Subtitle A, par. 4116(c), (d), 104 Stat. 517, applicable as provided by par. 1020 of such Act, which appears as 33 USCS par. 2701 note).

6. *Associated Press*, Sept. 18, 1999, in *The Olympian* and various papers.

7. *Associated Press*, Oct. 14, 1999, *The Olympian*, etc.

8. *Associated Press*, Dec. 7, 1999, *The Olympian*, etc.

Chapter 11

1. U.S. District Court for the District of Oregon, Civil number 89-609-RE: Sause Brothers Ocean Towing, a corporation, as Owner/Charterer of the T/V *Ocean Service*, Plaintiff, in a cause for exoneration from or limitation of liability, and related cases (Opinion filed Jan. 24, 1991).

Chapter 12

1. *Spill News*, Washington State Dept. of Ecology, Sept. 1993, Vol. 2, No. 1.

2. *Associated Press*, published in *The Olympian*, Oct. 15, 1994, and other newspapers.

3. Makah is not the name preferred by Neah Bay folk, but rather the name used by Clallam people to the east when asked. "People of the Cape," *Kwidishda-akh* is their own name, according to Vince Cooke, and their language.

4. *Shipwrecks off Juan de Fuca*, by James A. Gibbs (Binfords & Mort, 1968) pages 63-65.

5. State Labor and Industry insurance regulations exclude worker coverage for volunteers under age 18.

6. *Bobot'klid*, tribal word for non-tribal people, translates literally as "house that walks on water," referring presumably to the sailing ships that brought early explorers and traders to the Pacific Northwest coast.

Chapter 13

1. *Nestucca Oil Spill OSC Report*, Washington State Dept. of Ecology, 1989, previously cited.

2. *The Nestucca Oil Spill*, Preliminary Evaluation of Impacts on the West Coast of Vancouver Island, prepared for Environment Canada and British Columbia Ministry of Environment, March, 1989, by W. Duval, S. Hopkinson, R. Olmsted and R. Kashino, ESL Environmental Sciences Limited, Vancouver, B.C.

3. *Nestucca Oil Spill Report*, Canadian Coast Guard, June, 1989, previously cited.

4. *The News Tribune*, Tacoma, WA, Feb. 18, 1999.

5. *Spill Scene*, Vol. 3, No. 3, Summer 1999, Washington State Dept. of Ecology.

Chapter 14

1. *Washington Administrative Code*, WAC, Chapter 173-183.

2. *Nestucca Oil Spill Restoration Plan*, by Jeffrey J. Momot, April 1995, U.S. Fish & Wildlife Service, North Pacific Coast Ecoregion, Western Washington Office, Olympia, Washington 98502.

3. *Revised Code of Washington*, RCW 82.23B, Oil Spill Response Tax (assessed at 2 cents a 42-gallon barrel for spill response and 3 cents for administration).

4. *Washington Administrative Code*, WAC, Chapter 317.

5. *Spill Scene*, Spring 1999, Washington State Dept. of Ecology.

6. Final Report of the States/British Columbia Oil Spill Task Force, October 1990.

7. *Report to the Premier on Oil Transportation and Oil Spills*, by David Anderson, Special Advisor, November, 1989.

8. *Marine Oil Transportation Systems Evaluation of Environmental Risk*, D.F. Dickens Associates, Ltd., August 1990.

9. *Report of the Tanker Safety Group*, U.S. Coast Guard, 1989.

10. States/British Columbia Oil Spill Task Force, Jean Cameron, executive coordinator, 811 SW Sixth Avenue, Portland, Oregon 97204-1390, telephone (503) 229-5720, fax (503) 229-6124.

Chapter 15

1. Emergency Management, formerly a division of the Department of Community Development, became part of new Department of Community, Trade and Economic Development in March of 1994.

2. States/British Columbia Oil Spill Task Force *Annual Report, 1992-1993*, tells of a plan for task force members to adopt a uniform oil spill reporting number, 1-800-OILS-911, to "ring in the appropriate state Emergency Management Division based on the area code from which the call originates. The goal is to streamline spill reporting so that anyone...from California to British Columbia (and Alaska) has to know only the...one state reporting number."

3. *Geographic Response Plan*, Strait of Juan de Fuca, developed jointly by the Washington State Department of Ecology and the U.S. Coast Guard, 1993. Participants included local representatives, industry and response contractors, federal and state department, bureau and agency representatives. For more information contact Dick Logan or Paul Heimowitz, Washington State Department of Ecology, Olympia, Washington 98504-7600; telephone (360)407-6971 or (360)407-6972 or (360)407-6000.

4. Responsible officer for the Makah Tribe is Vince Cooke, phone (360) 645-2201. He is identified along with the tribal management requirement on a final overlay which appears in the GRP workbook, but was omitted from this illustration to avoid visual confusion.

5. Rob Whitlam, NRDA representative from Washington Archaeology and Historic Preservation, is the contact person, phone (360) 753-4011, and again that final overlay was omitted to avoid complicating the geographic view.

6. Leni Oman has since taken a different position within the department.

7. *Abstract, Washington's Marine Oil Spill Compensation Schedule-- Simplified Resource Damage Assessment*, by Laura Geselbracht and Richard Logan, Washington State Dept. of Ecology, PO Box 47600, Olympia, Washington, 98504-7600.

8. *Spill News*, January, 1994, a publication of the Washington State Dept. of Ecology, PO Box 47701, Olympia, Washington 98504-7600.

9. *Associated Press*, Nov. 3, 1999.

10. *Associated Press*, Oct. 27, 1999, *The Olympian*, etc.

11. *Associated Press*, Oct. 30, 1999.

12. *Associated Press*, Dec. 13,1999.

13. *The Olympian*, July 29, 1999.

14. *The Bellingham Herald*, June 11, 1999, et seq. This quantity of gasoline was more than the approximately quarter million gallons of crude or heavy oil spilled by either the *Nestucca* or *Arco Anchorage*.

15. *The Olympian*, Dec. 6 and 8,1999.

16. *The Olympian*, July 29, 1999.

17. *The Olympian*, July 29, 1999.

18. *The Olympian*, Dec. 8, 1999.

19. *Seattle Times*, Dec. 4, 1999.

20. *Seattle Post-Intelligencer*, Dec. 4, 1999.

Chapter 16

1. *Handbook of North American Indians, Volume 7*; William C. Sturtevant, general editor; Wayne Suttles, volume editor; Washington, D.C., Copyright 1990, Smithsonian Institution. Sold by U.S. Govt. Printing Office. References to other native peoples, such as the Quillayute and Nuu-Chah-Nulth, are also dependent in part on background information from this volume.

2. *Tillamook Indians of the Oregon Coast*, by John Sauter and Bruce Johnson (Binfords & Mort, 1974; Portland, Oregon).

3. *Handbook NA Indians, Vol. 7*, op. cit., page 567, synonymy: "Nestucca...Synonyms include Salmon River."

4. Other references: *Oregon Geographic Names*, Lewis A. McArthur (4th ed.); *Bibliography of the Salishan Languages*, James C. Pilling, 1893, Smithsonian Inst. Bureau of Ethnology (Bulletin 16); *Handbook of American Indians North of Mexico*, Frederick W. Hodge, editor, Pageant Books, New York, 1959, reprinted: Smithsonian Inst., Bureau of Ethnology Bull. 30; *Dictionary of Puget Salish*, Thom Hess, University of Washington Press, Seattle, 1976; *A Dictionary of the Niskwalli Indian Language*, George Gibbs, 1877, Nisqually Tribal Library, 4820 She-Nah-Num Dr. SE, Olympia, Washington 98503.

5. State of Washington, et al., v. Washington State Commercial Passenger Fishing Vessel Association, et al., U.S. Supreme Court, July 2, 1979.

Index

about the author

DAVID C. WEBSTER has been a Washington State resident more than 50 years, and even longer attuned to wildlife and things natural. He was born in Bend, Oregon, 1924. Four years later, his family moved to Sitka, Alaska, with forest out the back door and ocean beyond the front lawn. In 1940, they moved to California, where Dave finished high school and earned a BA in 1946 at San Jose State College.

Journalism called, the family had returned to Alaska, and he found a reporting job in Juneau. News work followed in Wenatchee, Spokane and Aberdeen, where he became city editor for six years. A new job, bailiff with the Washington State Supreme Court with PR aspects, began in 1964 and continued until retirement in 1987.

In addition to a lifetime as wordsmith, Mr. Webster has been an active church member and choral musician since childhood. Married since 1949, he and Mary have three children: John in Spokane, Mark in Gig Harbor and Constance Anderson of Eugene, Oregon. They, in turn, have seven fine grandchildren.

David and Mary have called Olympia home since 1965.

Also available from
Malamalama Press

The Imprisoned Splendor:
Discovering Your Spiritual Self

by Robert A. Wasner

"...filled with spiritual nuggets that open the heart
to the essence of our being. I highly recommend it."
- Gerald G. Jampolsky, M.D. author,
Love is Letting Go of Fear

* * *

Artful in its simplicity, *The Imprisoned Splendor*
is a clear, practical guide to spiritual living and growth.
Compared to the classic, *Illusions* by Richard Bach,
for the "same profound quality" of always having
"something to fit whatever difficulty one was facing."

* * *

"A gentle key that unlocks the divine...captures the essence
of spiritual laws and precepts. Wasner grabs powerful
examples from scripture and life,
then fashions a new foundation for living."
- *The Book Reader*, Independent Review of Books

* * *

Readers continually remark that *this*
is the book they refer back to again and again --
always heartfelt and always thought-provoking.

* * *

"Helped open me up even more to the wonder
and abundance of the universe."
- Wally Amos

Also available from
Malamalama Press

Dangerous Vision
by Gay Burk

Is Kym Hallucinating? Find out in this provocative suspense thriller with a paranormal edge. Haunting visions come unbidden to young Kym Nicholl after her arrival in England, fresh from America. Amidst a budding romance, the questions only deepen and suspicion soon blankets the charming village of Torping-Strand. While tensions mount around her, Kym holds fast to her secret knowledge -- that her visions have the power to heal a dark and hidden past.

* * *

GAY BURK, a resident of Hawaii since 1950, has decades of experience as a world traveler and writer. Her travel features and metaphysical articles have appeared in many magazines, including *National Geographic*, *Touring Times* and *Science of Mind*. She has also published both adult and children's stories, and has written books of fiction, biography and travel.

Dangerous Vision

____ copies @ $12.95 each

The Imprisoned Splendor

____ copies @ $10.00 each

Nestucca: An Oil Spill Turns Creative

____ copies @ $16.95 each

Shipping: Include $2.00 for first book;
$1.00 for each additional book ordered.

Washington residents add sales tax.

Total Enclosed : $ _____

Please send check or money order
in U.S. funds to:

Malamalama Press
PO Box 7185
Olympia, WA 98507-7185

Name _____

Address _____

City _____

State/Province _____ Zip Code _____

Phone Number _____

thank you for your order